Novice's Handbook to Chicken Keeping

Practical Tips and Proven Strategies for Raising Healthy Backyard Chickens

Nicole K. Murphy

Copyright

Table of content

4

Dedication

To those who find joy in the gentle clucks, and to the novices venturing into the world of feathered companionship. This handbook is dedicated to your journey of raising happy and healthy backyard chickens.

With warmth,
Nicole K. Murphy

About the Author

Nicole K. Murphy, the author of **"Novice's Handbook to Chicken Keeping,"** is a passionate advocate for the harmonious coexistence of humans and their feathered friends. With a background in animal science and a deep-rooted love for sustainable living, Nicole brings a wealth of knowledge and hands-on experience to the world of backyard chicken keeping.Nicole's journey into the realm of poultry husbandry began as a personal quest for fresh, wholesome eggs and a desire to cultivate a more connected and sustainable lifestyle. Her practical approach to chicken keeping is grounded in years of dedicated research, trial, and error, making her a trusted guide for novices embarking on their poultry adventures.A devoted advocate for ethical and compassionate animal care, Nicole believes that raising healthy backyard chickens is not just a practical endeavor but a rewarding and fulfilling lifestyle. Her commitment to sharing proven strategies and practical tips in this handbook reflects a desire to empower novices with the knowledge needed to create thriving flocks and nurturing environments.**Nicole K. Murphy** invites readers to join her in this exploration of the intricate world of chicken keeping, where practical wisdom meets the joy of building meaningful connections with our feathered companions.

Introduction

Welcome to the book "Novice's Handbook to Chicken Keeping: Practical Tips and Proven Strategies for Raising Healthy Backyard Chickens." If you've ever been attracted to rearing hens in your backyard, this booklet is a must-have companion on your way to becoming a self-assured and *empathetic chicken keeper.*

For those who are new to the world of poultry husbandry, we give a detailed introduction to the art and science of chicken keeping in the pages that follow. This manual is meant to provide you with the knowledge and valuable suggestions necessary to accomplish your aspirations of having fresh eggs delivered to your door, experiencing humorous behaviors from your feathery friends, or simply having a meaningful connection with them.Every chapter gets you closer to establishing a calm environment for flock and keeper alike, from selecting the breeds of hens that best fit your tastes and available space to constructing the perfect coop that puts the health of your flock first. We address basic equipment, and best methods for feeding your chickens, and even help you grasp local rules to make sure your backyard hens fit in with the community.However,this booklet gives a compassionate approach to keeping hens and goes beyond plain teaching. We dig into the complexities of hen behavior, understand the meaning behind their clucks and coos, and give assistance in designing exciting surroundings that increase their well-being. You will learn how to establish a symbiotic connection with your hens, which is much more than simply basic care and provides for a genuinely delightful experience.

Why would you want to avoid losing out on this manual?

★ **You'll find not only valuable tips but also a philosophy of respect and care for your feathered companions in these pages. You'll learn the self-assurance essential to handle difficulties, make good decisions, and play an active role in the greater community of hen lovers.**

★ **You won't be able to experience the enjoyment that comes with owning a growing backyard flock that adds value to your life if you miss this. You will miss the satisfaction of picking**

your fresh eggs, the delight of watching your chickens' personalities, and the comfort that comes with being an educated and responsible chicken keeper.

So come along on this amazing adventure with us, dear reader. Your ticket to a world in which pleasing and healthy backyard hens are not just a dream but a lovely reality is **"Novice's Handbook to Chicken Keeping"**. Don't pass up these precious recommendations that will assure the success of your chicken-keeping attempts. With this manual in hand, you're set for a pleasant and fulfilling experience in the fascinating world of backyard chicken keeping. Your future flock is waiting for you.

Chapter 1

How to Begin Raising Chickens

Taking on the work of rearing chickens demands meticulous preparation, backed by a full knowledge of the underlying ideas. This first chapter functions as a road map for prospective hen keepers, covering the key chores required to establish and maintain a thriving backyard flock. The cautious choice of a chicken breed that suits the keeper's preferences and the accessible space is the beginning point of this quest. Certain breeds have distinct qualities, such as temperament, size, and desire to produce eggs. By carefully examining these features, one may create a flock that complements individual interests and develops well within the assigned area's geographical restrictions.Building a chicken coop is one of the main architectural parts of rearing hens. The intricacies of coop design are explored in length in this subchapter, with an emphasis on vital components such as optimal ventilation, properly made nesting boxes, and well-placed roosts. Beyond the essential demands of shelter, a well-designed coop serves as the basis for the flock's overall well-being, assuring them protection, comfort, and excellent health.Purchasing the essential tools and equipment is also vital when establishing a chicken-keeping company. This sub-chapter meticulously discusses the fundamental components required to keep a clean and productive henhouse, from properly built feeders and waterers to the choosing of acceptable bedding materials. Gaining a good knowledge of this equipment is crucial to the flock's daily care and control, which in turn enhances their overall health. A key and sometimes ignored part of the planning stage is having a solid grasp of zoning rules and local bylaws. Knowing the local laws, zoning rules, and compliance requirements is vital when navigating the legal environment of urban or rural chicken farming. This sub-chapter discusses the nuances of these aspects, highlighting how vital it is to coordinate one's chicken-keeping initiatives with the current regulatory frameworks. Early compliance with regulatory criteria enables a smooth incorporation of chicken agribusiness into the greater community. Prospective chicken keepers need to realize that there are key aspects that affect the path of a prosperous venture, hidden beneath the lovable clucks and noisy bustle. A backyard flock may grow when certain circumstances are fulfilled, including careful breed selection, well-thought-out coop design, procurement of appropriate equipment, and a complete grasp of local rules. Therefore, this chapter works as a full introduction, setting the scene for the range of chores and advantages that are characteristic of the chicken-keeping world.

A complete guide to picking the perfect breed of chicken for you

In the past, I used to ask myself, "What breed(s) are best for me and my situation, location, etc.?" when I first began raising hens. Now that I've had a few years of experience and research under my belt, I've opted to put up a guide to aid individuals who are new to raising hens making the decision much easier. I'll go over various things and build a code for every breed below.

Section 1: Objective

The first thing to think about is your unique aims for rearing chickens. Eggs? Meat? Just for fun? Perhaps every one of these? All these criteria may be supplied by hens, but the degree of benefit in each area varies considerably on the breed you pick. Your selection of birds should be highly affected by the primary function, or the characteristic, that each breed is recognized for.

The original code and category

Eggs (E): Which came first, the eggs or the chicken? That's a topic for another chapter. Naturally, however, it may be the most usual non-industrial function for these species. Many of the contemporary kinds of chicken were created for their delectable, healthful **"butt nuggets,"** with some of them laying an egg virtually every day of the year. A distinct laying setting, such as a nest box, should be offered for laying hens; ideally, there should be one box for every three birds, if not more. While eggs may be found in a range of sizes and shapes, pick one of these breeds if you want your birds to lay eggs by themselves.

Meat(M): Is there anything that tastes more like chicken than chicken? Nothing! Meat birds may create vast quantities of tasty meat when given the correct care. Since these creatures normally mature incredibly rapidly, rearing them should be done as inexpensively and effectively as possible. If your primary interest is meat, a broiler or fryer bird breed might be appropriate for you, even if some of them have difficulties reproducing and may not be especially good layers.

Show(S): I've used the word "show" to apply to both animals on exhibit and breeds maintained purely for enjoyment. Both giant and bantam (small) chicken breeds are primarily grown for looks rather than usage in agriculture. These breeds may not be the plumpest or the most prolific egg producers, but they are nonetheless attractive if you're searching for a striking or plain cute breed.

Dual-Purpose (D): This category comprises a surprisingly high number of birds that excel in both the nest and the eating area. Breeds with multiple functions are famous for their prowess in either laying and eating, laying and exhibiting, or all three! A dual-purpose animal might be the appropriate breed for you if you're seeking for a flock that can supply eggs, meat, and good appearance.

Section two: Dimensions

It matters, at least with chickens, I hate to break that to you. A larger chicken requires more **area**. If your space is confined to a few square feet, get **Bantams** instead of **Brahmas**, unless the breed is better acclimated to confinement. That is, of course, unless the breed is **Brahma bantam**. There are many various sizes of chickens, from the small Serama to the appropriately named Jersey Giant. Just as some breeds are incompatible with specific housing, places, and equipment, elephants cannot employ cat flaps. Before making a decision, you have to examine your size limits.

Bantams: The name "bantam" refers to a type of chicken that is naturally relatively little. While some breeds are "true bantams," meaning there is no bigger counterpart, other larger breeds have bantam counterparts. These small birds are charming, but don't expect enormous eggs from them! According to some literature, 8" by 8" is adequate, but I feel that each bird requires a minimum of 2 square feet of territory. (Since chickens are sociable creatures and you should never keep fewer than three together, don't expect a two-square-foot coop with a single hen to act as a hamster cage.)

Small: Bantams are considered as enormous birds, nevertheless, there will always be the smallest one in the bunch. able to lay visible (although not massive) eggs, but not as strong as bigger breeds when it comes to punch, peck, or kick. A minimum of 3–4 square feet should be allotted for every animal.

Medium: most layers should be located here, in my view. With enough size and appetite to create gigantic, gorgeous eggs, but not so enormous that they require a lot of area or a large body. For each bird, I'd advise utilizing at least 4 square feet.

Enormous: an animal of this size may be made that serves two functions well since its body can maintain parts that are massive enough to contain large eggs and enough meat to feed a small number of people. Because they were initially produced in little communities for both meat and eggs, heritage breeds are frequently enormous. Per bird, 4-6 square feet.

Extremely large: well, here you have it if, like me, you're either able or mad enough to pick chickens the size of small children. At 2'6" now, my Brahma rooster is nowhere near the biggest example. While some of the breeds in this group are the friendliest around, keep in mind that the roosters may be very aggressive and huge enough that they may not always see humans as a danger. These large creatures may occupy almost 8 square feet of chicken.

Temperature suitable for chickens

Certain breeds are just not appropriate for severe regions. certain may become overly hot or cold. As a consequence of their origins, breeds just adapt to varied environments. If you dwell somewhere that is less tolerant, there might be certain breeds you should keep away from.These are frequently plumper or feather-footed, and they are cold-resistant.. Although no breed of chicken is ideal for the Arctic climate year-round, cold-hardy varieties are more resilient to severe winters than other varieties. This might lead to overheating in really warm places, so this might not be the best option if you encounter both extremes.

Mild/average

These birds do best in environments that are in the **Goldilocks zone**—that is, **neither too hot nor too cold.** They won't be able to withstand either extreme for extended periods, but they'll be great if you just have a few warm or cold weeks each year. It is often the case that breeds that are milder will also work well in these situations, as will all-around/heat/cold hardy birds.

Heat Environment

These birds, which mainly originate from tropical or Mediterranean breeds, are less prone to overheat than other breeds. Compared to their northern counterparts, they are frequently leaner and more nimble, with fewer and looser feathering to help cool off on a hot summer day.All-rounders are breeds that thrive well in a range of weather situations, including hot and cold. One of these breeds will keep your flock sound all year round if you dwell in an area with varied seasons and extended periods of both heat and cold.**Brahmas are good all-rounders because, in addition to their apparent cold adaptations, they flourish in hot temperatures.**

Laying eggs

If your pet prepared your breakfast, wouldn't that be amazing? For this, hens are great since they can produce food and yet be alive. The breed will also determine how much of a breakfast you may expect; some kinds lay one egg every day, while others only lay a handful each year. It is crucial to balance the amount of eggs you have with the number of hens you need to lay each day or each week. If you simply need a few eggs for the weekend, you don't need 20 Leghorns, and if you want to feed your whole neighborhood with eggs, you don't need an Ayem Cemani. This data only covers the hen's first year of laying; as she gets older, it gradually declines.

0–50: Breeds with an egg output this low are frequently crossed with wild jungle birds more recently or are produced exclusively for the show. The spring and summer months will see the bulk of these eggs deposited.

50–100: near the lower end for chickens, but still a major step ahead over wild birds. A regular diet should contain calcium.

100–150: These hens should be given poultry mix, which includes layers of pellets for eggs, and regular food for chickens. You may expect a few eggs each week from them.

150–200: reliable layers that give an egg roughly every other day; layer food is advised to protect egg quality.

200–250: Let's get serious now! These are for you if you use a lot of eggs. It may be conceivable for you to begin selling depending on the numbers. The major source of nourishment should be layers or pellets.It doesn't get much better than **250–300:** An almost daily egg, so be prepared for them. Feed feed or pellets to layers to preserve body health and egg production.

300+ usually only possible with hybrid vigor, however, the leghorn is notable for being able to do this. These kinds of birds are raised commercially for their eggs in stores. Should be on layers of feed to avoid problems from laying so regularly; others think pellets are the best way to ensure that selective eating doesn't cause them to miss out on important nutrients from the diet.

Known for their use in industry, Lohmann Browns lay more than 300 eggs yearly.

Color of the egg

How many hues an egg may come in might astonish even the most expert chicken keeper. The list includes blues, browns, white, green, speckled, green with chocolate hues, and so on. I am aware of the enjoyment that may arise from gathering your morning's eggs in the dark, not knowing precisely what to expect, and then opening the coop to see a rainbow of colors so vivid you won't want to break them open. Pure breeds are distinguished by the hue they are bred to lay; others are valued for their brightness, depth, or speckles.

White: the original hue of eggs produced by wild jungle fowl, or the natural kind. While many people are used to white eggs, others in the UK, like myself, may find this a bit more unexpected. This gene is recessive against blue eggs; the result is a pearly white shell devoid of any of the pigmentations seen in brown or green eggs.

Brown: Brown eggs are more likely your regular supermarket kind if white eggs aren't. Initially, the genes determining the amount of a brown pigment, or paint, added to the shell were the same as those governing white eggs.

Cream/tinted: Several breeds that aren't expressly bred for eggs produce cream eggs, which are less pigmented than brown eggs. These eggs may be a variation from more normal colors due to bloom and calcium deposits, which may make them look pink under certain lighting settings.

Blue: Blue eggs are frequently the beginning step toward understanding that there are more egg colors than the ordinary person may know. Because blue eggs are dominant over white shells and lack brown pigmentation, if they were to crossbreed, the offspring would likewise lay blue eggs.

Russet: We are now going into the more odd colors. Russet eggs are speckled and commonly deposited by a few breeds; these are not to be confused with brown eggs, which have a dark brick-red color.

Chocolate: The most striking displays contain eggs that are as dark as dark chocolate. Even olive and russet eggs pale in contrast to these, the darkest colors possible for eggs. Their speckles are also prevalent.

Green: Maybe with some ham? The yolks should not be green, but the shells may have shades similar to those of fruits, grass, and foliage without being as dark as olive eggs. Only a small number of varieties—mostly variations of blue-laying breeds with a little amount of light color added—lay green eggs. Because the genes from brown layers are transferred to blue eggs, crosses like Easter eggers can produce green eggs. Perhaps the only real green layer chicken that is purebred is the Silverudd's Blue (Isbar) breed.

**Olive/moss**: Olive eggers" are a wonderful addition to any flock, and they are only generated by crossovers between blue and dark brown layers. As the name implies, the colors of the shells vary from olive to a dense covering of emerald-green moss. Their shells resemble those of dinosaurs (actually, they do, but you get the idea), sometimes inherited from their parents who laid chocolate.

Important Information for Novices Regarding Poultry Equipment

Tools and equipment used for hatching, brooding, housing, feeding, cleaning, and preserving optimal conditions in poultry farms are referred to as **poultry equipment.** Small-scale farmers need manual equipment to efficiently carry out their tasks, but at the commercial level of production, the vast quantities of birds will need more advanced machinery and skilled people to operate them. There are many options for equipment, depending on your production level.

Standards for Choosing Equipment for Poultry Farming

Bird Type: Depending on the breed and needs of the birds, you will need to employ varied equipment to raise each one.

region of Specialization: Because the poultry industry has a complex value chain, you should also take your region of poultry production into account when purchasing equipment. chicken farming includes the production of eggs, meat, and packaging, as well as chicken feed. Your equipment selection will depend on your position inside the value chain.

Space: One of the most significant factors when selecting poultry equipment is the amount of birds on your farm.

Price Comparison: The cost is still another vital aspect. There are numerous items to purchase, so decreasing expenses by a few naira on certain goods will decrease the overall cost.

Technical Know-How: It's vital to evaluate the technical know-how needs of the equipment you pick. Before purchase, it should be established who will use it and how.

Technology: When acquiring poultry equipment, technology is also vital. You already know that technology is continually evolving. As such, you need to make sure that the poultry equipment you are purchasing is technologically advanced by completing a comprehensive inquiry.

Typical Poultry Farming Tools and Their Applications

Incubator

The purpose of incubators is to guarantee the secure and fruitful hatching of eggs. An incubator is a device that mimics the incubation of birds by maintaining eggs at the proper humidity and temperature range, together with a rotating mechanism that allows the eggs to hatch. The hatching time of a chicken egg is 21 days. Incubators come in a variety of forms and sizes, with different capacities for eggs and working mechanisms.

Gas-powered poultry brooder

A brooder is a heated building used to keep chicks warm. Studies have shown that chicks exposed to low temperatures suffer from compromised immunological and digestive systems. Consequently, cold-stressed chicks develop less and become more vulnerable to illness. It is essential to provide the chick with some heat. A few brooders control the heat and maintain the ideal temperature.

Cages of Chickens

The kind of cage needed depends on several aspects, such as the kind of bird (layers work better in battery cages) and the production method (open or deep liter systems don't need cages). Cages exist in a variety of sizes and configurations, and their frames are supported by either iron or wood rods.

Broiler Battery Cages: Designed for hygienic and easy brooding in constrained land areas, broiler battery cages are semi-automatic types. They include all necessary water accessories and related equipment. All varieties of poultry chicks and adult birds in an all-in, all-out intensive management approach may be housed in the cages.

Layer Battery Cage: Chicks are housed in compartment units inside this sort of intensive poultry housing system designed for layers. The battery cage system gets its name from the way similar-looking cages are arranged in rows and columns. The flooring of the Layers battery cage slopes from back to front, setting it apart from most other cages. This is so that the eggs may easily be collected by rolling them from the rear to the front of the cage.

Carry Crates

"Transport crates" are enclosed boxes meant to make transportation simpler. They may be incredibly convenient for transferring birds around, especially from farms to hatcheries.

Drinkers and Feeders

Birds' degree of productivity is largely determined by their continual availability of food and water. Farmer preferences for feeding and drinking equipment are also influenced by the production system. There are various types of feeders and drinkers; some are automated, while others are hanging. Chickens are fed from containers called poultry feeders. They preserve the hens' feed and provide them with a secure spot to eat without wasting any. The goal of breast drinkers is to benefit poultry birds by providing them access to safe drinking water. This form of drinker combines the nipple, clip, and drip cup into one drinker. Layers, breeders, and broilers may all use them. Bell Drinkers are used to guarantee that poultry birds, from day-old chicks to fully grown chickens, have access to adequate water when they are cared for utilizing a deep litter system.

8. Vaccine Supplies

A particular piece of equipment is utilized to supply farm animals with drugs and immunizations. In poultry farms, intravenous injectors are employed widely.

9. Debeaking Device

The process of shortening a bird's beak, particularly in layers and turkeys, is known as "debeaking" in the poultry industry for a variety of reasons. It's also known as conditioning or beak trimming. The infrared debeaking machine and the hot blade debeaking machine are the two kinds of debeaking equipment.

10. Equipment for Processing Poultry

Processing equipment is used in the poultry production value chain throughout the packing process. Mature chicken birds may be successfully killed, de-feathered, and packed with the use of specialized equipment.

11. Egg Containers

These are Equipment for rearing chickens.Egg cages are specific to layers and assist transfer and storage of chicken eggs.The appropriate equipment must be employed for a poultry firm to be successful. Purchasing equipment for chicken farming from trusted providers and making sure it is in great functioning condition are both vital.

Comprehending Local Laws Concerning the Raising of Backyard Chicks

Having hens in your garden is an enjoyable and rewarding way to live. However, you must research the activity's legality before beginning. Town-to-town differences exist in the legislation about chicken ownership. For example, limits on the number of chickens allowed per flock or family are uncommon in rural regions. Right to Farm statutes are in effect in all fifty states to shield homesteaders and farmers against lawsuits claiming they are causing nuisances. Unfortunately, raising chickens in suburbs and cities is sometimes subject to regulations and is not covered by these rights to farm. There might be differences in regulations regarding chicken keeping from block to block or neighborhood to neighborhood even inside municipal borders. Purchasing backyard chickens requires understanding the legality of the birds before making the actual purchase. A centralized database containing such information is still lacking, and you shouldn't rely solely on search engines to locate the information you need because laws and guidelines referenced in newspaper articles and blog posts can change rapidly, particularly in light of the recent push for legislation about poultry. Here are a few instances of the constantly evolving legislation found in the nation's largest cities and metropolitan areas.

New York City Laws Regarding the Raising of Backyard Chickens

It is permissible to raise hens in any of the five boroughs of New York City; however, other poultry and birds, including turkeys, ducks, and geese, are not allowed. The city takes neighbor complaints seriously and demands that you abstain from producing *"nuisance conditions" (noise or smell).*

Texas rules concerning backyard chicken farming

A Texas congressman introduced a measure in 2018 that would loosen regulations on backyard flocks. Historically, the state has had disparate rules regarding the keeping of backyard chickens; nevertheless, the measure seeks to standardize this area. As of this summer, the proposal, which was approved by the Texas Senate in the spring of 2019, is still pending a House vote.

New Jersey legislation concerning backyard chicken farming

In New Jersey, keeping hens in one's backyard has risen in popularity. As a consequence, municipal rules are continuously altering. For example, some localities have implemented ordinances addressing noise, property boundaries, and chicken cages.

Oregon rules concerning backyard chicken farming

If you maintain more than three hens or ducks, you must get a permit in Portland, Oregon. For three or fewer backyard hens, Portland does not need a permit, nevertheless, owners must maintain an absorbent ground cover and keep their chicken cage at least 15 feet away from any nearby residential structures.Michigan legislation concerning backyard chicken farmingWhile most communities in the state allow chicken keeping, Detroit, Michigan, does not. As interest in backyard chickens develops, Detroit's tough ban on them has spurred controversy, with many pushing for more relaxed rules.

Illinois legislation concerning backyard chicken farming

In backyard settings, Chicago permits both hens and roosters (noise issues aside). Chicago, like many other jurisdictions, restricts the rearing of animals in residential districts for slaughter, whereas pets and hens that give eggs are permitted.

How to identify and understand the laws in your region

Get in touch with your town or city hall immediately to learn about and comprehend your local regulations. Administrators can explain to you precisely what is and isn't allowed in your community at this point. After obtaining this permission, you need to think about having a conversation with your neighbors. Although you don't have to ask, you should consider the potential benefits and effects that owning chickens may have on them. Make sure to introduce yourself and tell them why you raise chickens, along with the exciting potential. It might be prudent to sometimes give them a half dozen eggs to assist in ensuring their support for your new venture.

WAYS TO PREVENT ANNOYANCE REPORTS

The majority of complaints about backyard chicken nuisances are related to *sound and smell problems*. As a group, hens smell significantly less than bigger animals, such as a backyard cows. There are restrictions on how many chickens you may have in your backyard in many places because if you possess a large flock of hens, their area may start to smell.Hens are often not noisy neighbors, but roosters are a different story. Male chickens called roosters are not legally required to rear hens. Even in the absence of a rooster to fertilize the eggs, hens will still lay eggs. A hen may squawk with pleasure when an egg is deposited, but most of the noise comes from the roosters. This call comes at a moment when even the noise of a city is at a pleasant ebb, resonating off buildings and skimming through open summer windows far before most people are ready to start their day. Consequently, roosters arc now illegal in many highly populated areas; in those that are not, like Chicago, you are at the whim of your neighbors' tolerance.

Putting Boosters in Your Stream

You should think about adding a rooster to your flock if both your town and your neighbors allow them. If you retain fewer than that, the roosters will start fighting over what they consider to be their area. It is advised that you keep one for every ten hens. A good rooster provides the priceless protection of his hens despite being loud. As the hens go about their daily business, roosters tend to spend much of their time with one eye fixed on the sky and the surroundings, guarding against predators and other birds of prey that may endanger their flock. He uses a succession of low murmurs to signal to his hens when it's safe to hide and when to come over and share a delicious worm.Additionally important if you want to incubate and hatch your eggs are roosters. An egg cannot grow into a chick without the fertilization that roosters supply; without one, it will only create breakfast.Whether or not a home has a rooster, raising chickens is becoming more and more popular among urban and suburban households that previously felt they were immune to the joys of home cooking.Consider launching a petition to have backyard birds allowed in your neighborhood if the practice is currently prohibited by local legislation. You can practically taste the rewards of your effort and take control of your food production with a small amount of outside space.

.

Chapter 2

How to Make Chicken Feed

Healthy chickens require a well-balanced poultry feed. A subset of free-range chickens supplements their natural diet by ingesting nutrient-rich poultry feed. The most critical thing you can feed your flock when it's maintained in a coop and run is **high-quality feed**. Is it viable to create feed for your chickens? How should one combine grains to attain a nutritious balance? Continue reading to discover how. Look at the formulation necessary for laying hens before you start buying bags of bulk feed and nutritional additives. The basic purpose of feed mixing is to deliver the finest possible nutrition in a pleasant combination. If your hens don't appreciate the pricy grains you've thrown in, there's no point!

What Do Chickens Need to Eat to Maintain Their Nutrition?

Poultry Feed

Chickens have various nutritional needs that must be supplied by their feed, just like any other animal. Proteins, lipids, and carbs come together in a balanced mix to make the nutrients available to the chicken's body. The second crucial item that is required in every diet is **_water_**. There is a tag on the packaging of commercial chicken feed that displays the nutritional contents in percentages. A standard layer of chicken feed has a protein value of 16 to 18 percent. The proportion of protein that is available for digestion differs between grains. Mixing your feed may be done with different grains. Consider picking grains that are organic, non-GMO, free of soy, free of corn, or both. Make sure the protein composition of a chicken feed ration maintains between 16–18% when replacing feeds. A bag of chicken feed already has the formulation done for you. The feed firm determined the quantities using the parameters of a normal chicken. Making your chicken feed from home using a proven formula or recipe will ensure that the nutrients are balanced and that your birds are receiving the proper quantities of each.

Percentages of chicken rations utilizing bulk grains and nutrients

30% corn (I prefer to use cracked, but the whole is OK too)

30% wheat (cracked wheat is my favorite variety).

Twenty percent dried peas

Ten percent oats

Fish meal at 8%

Use 2% Nutri-Balancer or kelp powder to ensure you're receiving adequate vitamins and minerals.

Preparing your Poultry Feed

How to Prepare Chicken Meal

Purchasing enormous sacks of each component from a grain supplier or feed dealer is the optimum technique to mix poultry feed if you have a large flock of laying hens. Finding a source for the components may involve some study and research, but it shouldn't be too difficult to obtain them. **Grain storage** is the next item to take care of. **Big metal bins or trashcans** with well-fitting lids keep the grains dust-free, dry, and protected from insects and rodents. It's crucial to forecast how much feed you'll need each month. If new grains lose their freshness, keeping them for more than a few weeks may end up costing you money. Alternatively, you may buy particular components in reduced amounts and manufacture your chicken feed instead of creating it altogether from grain. Whole grain bags weighing five pounds may be purchased by making an online order. This is an example of a recipe that you may use to generate roughly 17 pounds of layer feed. This may be sufficient to feed your backyard flock for a few weeks if it is modest.

Recipe for Small Batch of Chicken Feed

5 pounds. kernels or broken kernels

5 pounds. wheat

3.5 pounds. stale peas

1.7 pounds. cereals

One and a half pounds. fish meal

Five ounces (0.34 pounds) of Nutri-Balancer or Kelp powder, for appropriate intake of vitamins and minerals

Vitamin C for Hens

Grit for Chickens, Providing Appropriate Additions to The Herd

Two additional food ingredients that are typically offered free choice or added to the diet are **calcium and grit.** Calcium is important for the production of strong egg shells. The two most prevalent techniques for delivering calcium to the flock are adding oyster shells or recycling spent egg shells from the flock and giving them back to the chickens.For poultry, grit is small particles of ground-up dirt and gravel that the birds naturally accumulate when they are pecking the ground. We typically add it to the diet-free option to guarantee the chickens have enough as it is vital for good digestion. Grit gets trapped in the bird's gizzard, where it assists in breaking down harder foods like grain and plant stems. Reduced grit levels in chickens could lead to sour or damaged crops.Grubs, mealworms, and black oil sunflower seeds are wonderful sources of additional nutrition that the flock typically considers as rewards. These meals offer your hens a boost of protein, lipids, and vitamins, and they will also make them exceedingly healthy.

Prebiotics

Probiotic foods are regularly referenced in talks on diets for people and animals. Foods rich in probiotics boost the absorption of nutrients in the gut. Probiotics may be bought in powdered form, but creating your own is a straightforward procedure. Regular addition of probiotics to the chicken's diet may be done in two straightforward ways: **raw apple cider vinegar and fermented chicken feed.** You have all the elements required to produce a fermented feed when you mix your grains to make a homemade chicken feed. After only a few days of fermentation, whole grains are filled with beneficial microbes and have enhanced nutrient availability! Creating your chicken feed with the components you chose is more than merely a do-it-yourself effort. You are ensuring that the fresh, high-quality components of well-balanced diet are being provided to your flock. Which types of things have you utilized in your chicken feed? Has your flock experienced any ingredient failures?

These are food for thoughts,when preparing your next chicken food.

Advice for Effective Free-Ranging of Chickens

Imagine having chickens that are happier, healthier, and able to produce better-quality eggs because they have the freedom to walk around and examine their environment. With free-ranging chickens, it is doable! This topic will teach you how to offer your free-ranging flock the optimum blend of protection and freedom, so they may develop and flourish in their natural habitat.

A Brief Synopsis

Give your hens the freedom to move around and receive the advantages of happier, healthier birds and better-quality eggs!
Provide shelter for your flock's safety while securing them with robust fences and other security measures.To ensure

a favorable experience with free-ranging, predators must be caught, watched, regulated, and supervised.

Advantages Of Having Free-Ranging Hens

Free-range chickens that seek food on their own Several advantages come with owning free-range chickens as compared to limited ones. Let your chickens range free and you'll witness increases in their general health, happiness, and egg quality, to mention a few. However, how do these benefits arise? The natural behaviors and nourishment of free-range chickens contain the key to the answers, considerably boosting their overall well-being.To dive deeper into these benefits, let us analyze how allowing chickens to wander freely leads to happier, healthier hens and better-quality eggs.

Better-Fortified Chickens

Access to a diversified diet of plants and insects is one of the main benefits of keeping free-ranging chickens. These organic food sources offer critical vitamins and minerals, resulting in healthier chickens and better-quality output.

Free-range chickens produce meat and eggs with more protein content, lower fat content, and higher amounts of iron, zinc, and vitamins A and E owing to their nutrient-rich diet. As a consequence, your chickens are healthier and produce more delectable and nutritious meat and eggs.Free-ranging chickens receive more sunlight and are more active in addition to having a variety of diets. Their overall well-being and physical health are kept by their active lifestyle. You may expect a flock that is more robust and productive with healthier chickens.

More Joyful Avians

Every backyard chicken owner wishes to have a healthy, satisfied flock of free-range backyard chickens. Free-ranging allows hens, even those bred outdoors, to display their intrinsic behaviors, resulting in satisfied birds that consume less chicken feed. Among the numerous fun activities that free-range chickens may indulge in are foraging, dust bathing, and roosting in trees. These behaviors are known to be more prevalent in free-range hens than in loose birds.In addition to offering a cool escape from the summer heat and a warm sanctuary from the winter cold, trees, and shrubs also assist in producing healthier chicken eggs. You can ensure a healthier and more content flock of chickens, which translates to higher overall health and increased egg production, by allowing them to show their natural behaviors.

Better-Quality Eggs

There's no doubting the superior quality and taste of free-range eggs. A free-range egg has a substantially thicker and larger white, and the yolk has a richer, deeper golden tone. For those who maintain chickens, free-range eggs are a premium alternative because of their unique traits.These quality eggs are a product of the nutrient-rich diet of free-range chickens. Less cholesterol, less saturated fat, more vitamins, and more omega-3 can be found in free-range eggs. Because free-range chickens have a more natural diet and manner of life, they also tend to be bigger.Remember that by eating weeds, bugs (even unpleasant ticks), and worms, your flock is obtaining some pretty great nutrients directly from the environment.The difficulty of free-ranging is worth it when you can enjoy a more tasty and healthful breakfast with these superb eggs.

How to Get Your Yard Ready for Free-Ranging

Chickens running freely and foraging

Make sure your yard is safe and pleasant for your hens before letting them run free. Establishing the finest possible free-ranging environment for your flock requires reinforcing the borders, providing shelter, and enriching the surroundings. You can make sure your chickens have a safe and stimulating environment in which to explore and grow by following these guidelines.Let's take a closer look at safeguarding the area, offering your free-range chickens shelter, and enhancing their environment.

Safe Enclosure

Establishing a free-range setting for your hens should emphasize keeping them secure from predators. You may help keep predators away and your chickens safe by making sure the boundary of your yard is firmly fenced. Hardware cloth, chicken wire, poultry netting, and galvanized fence are some types of fencing materials. Selecting durable and trustworthy fence materials is vital for securing your flock. It is important to utilize extra security measures in addition to fence materials. To deter predators from digging, keep the grass cut around the perimeter, cover up any gaps in the fence, and bury wire mesh. Installing an electric fence around the perimeter to fend off larger animals is a wonderful option for greater security.

Giving Refuge

For hens that are permitted to range freely, the shelter is crucial because it offers shade from the sun, protection from predators, and shelter from severe weather. Natural hiding places and critical shade and shelter for your chickens may be found in trees, shrubs, and other vegetation. Your chickens will be safer in these organic shelters, and they will also have a more exciting space to explore. It's crucial to offer your chickens lots of cover in their free-range area so they have a safe and pleasant place to move around. Their satisfaction and overall well-being will grow as a consequence, making the flock healthier and more productive.

Improving the surroundings

Free-range chickens require an environment that is rich in nutrients for them to be healthy. You may provide your chickens with a more fascinating environment by growing natural grass, dispersing seeds, and adding stones and decomposing wood. These additions attract bugs and worms, providing your chickens with a more diversified diet and enhancing their overall health.Your hens will have a more enjoyable and exciting area to explore if you make the surroundings more interesting. They will be delighted by this, which will boost their overall joy and well-being.

Crucial Strategies for Managing Free-Range Chickens

Tick-foraging chickens

A free-ranging flock needs to be managed effectively, which entails supervision, training, monitoring, and predator management. You and your feathered friends will have a more joyful time with your free-ranging hens if you can assure their safety and well-being by adopting some key habits.This portion will explain how to raise chickens so that their experience with free-ranging is both safe and beneficial. This entails teaching the birds, keeping an eye on and regulating their behaviors, and managing predators.

Educating Your Chickens

For your flock's management and the preservation of your hens, training is necessary. A marker signal and an incentive, such as mealworms or cracked corn, can assist you in educating your hens to react when called. If required, this skill will allow you to fast and successfully gather your chicks.

It's vital to coop train your chickens so that they return to their coop at dusk in addition to learning to come when called. This will ensure that they return to their sanctuary every night, minimizing the likelihood of predation and other hazards.

Observation and Guidance

The security and welfare that come with regular supervision are experienced by monitored free-range chickens. By keeping predators away and reacting immediately to any accidents or attacks, you can make sure they are safe. Supervised free-ranging also helps you to spend quality time with your feathered buddies, enhancing the relationship between you and your flock.

Larger flocks would require additional staff for oversight. Just before nightfall is the greatest time for supervised free-ranging. ***Dogs and cats*** are examples of home pets that could be handy for keeping an eye on and maintaining your free-range poultry. They can keep your chickens safe and secure when they are permitted to range free by helping to fight off predators and notifying you of any risks.

Handling Potential Hunters

One of the most critical components of keeping free-range hens is controlling predators. Having a rooster in your flock is an excellent strategy to keep your hens secure and function as a warning mechanism for any prospective threat. Because roosters are attentive and will raise the alarm if they perceive danger, you and your flock may take immediate action to limit any harm. Having a rooster is not enough to keep predators out of your free-ranging area; you also need to put in place proper fencing and security measures. To prohibit predators from digging a route, this requires burying **wire mesh,** closing gaps in the fence, and deploying **high-quality fencing materials.**

How to Choose the Correct Breed for Free-Ranging

Chickens walk freely in the garden and forage

Selecting the proper breed for your flock's success and well-being while free-ranging is vital. A breed that is gentle, polite, adept at foraging, observant, and sensitive to its environment is wonderful. By picking the proper breed, you can be certain that your hens will prosper in their native environment and be more robust to any dangers. The temperament, ability to forage, and awareness of predators are all crucial characteristics to consider when selecting the appropriate breed of free-range chicken.

Characteristics

When picking a breed of chicken for free-ranging, temperament is a critical issue. This environment is best appropriate for calm, unruffled hens as they are more likely to stay close to home and acclimate well to being free-range. Though every chicken is different, in general, brave and social birds are better suited for being allowed outdoors. A calmer flock that is more likely to stay in their allocated free-ranging area will come from selecting a breed with a meek and friendly demeanor, which decreases the possibility of their straying or coming into touch with prospective predators.

Capacity for Foraging

Another essential criterion when selecting a breed for free-ranging is foraging capabilities. Foraging properly helps hens acquire food on their own and reduces the need for supplemental feed. Ameraucana, Ancona, Andalusian, Buckeye, Egyptian Fayoumi, Golden Comet, Hamburg, Old English Game, and Welsummer are among the breeds that are well-known for their propensity for foraging. You can be sure your hens will prosper in their free-range settings and stick to a healthier, more natural diet by picking a breed with exceptional foraging ability.

Awareness of Predators

Choosing a breed for free-ranging necessitates that the animal be aware of its predators. Alert and aware of their surroundings breeds will be better able to evade predators and maintain their safety and wellbeing. You can make sure your hens are more cautious and alert of any hazards by picking a breed that has great predator awareness. This will help them to be safe and secure when they are permitted to wander freely.

How To Balance Freedom And Safety When Free-Ranging

You must find the perfect balance between freedom and safety if you want your free-ranging flock to flourish. You may let your chickens wander about freely while decreasing the dangers associated with free-ranging by putting in place adequate security measures, providing shelter, and enhancing the environment. This portion will address how to establish the right balance between your free-range hens' freedom and safety, including how to utilize flexible fencing, the advantages of rotating pastures, and the benefits of using technology.

Adaptable Fencing Solutions

Managing the security and independence of your free-range hens relies greatly on your choice of adaptable fence systems. Fencing, either permanent or moveable, may restrict hens to a set area while still allowing them to roam and explore. Chickens that have had their wings clipped may be less inclined to fly over fences, but they may also be more sensitive to predators. You may obtain the optimum blend of protection and freedom by employing flexible fence options, which will enable your hens to travel freely while being guarded from any predators. Depending on your expectations, there are a variety of non-electric and electric portable fence solutions accessible online.

Turning Pastures

A fantastic approach to secure the security of your flock of free-ranging sheep and allow them access to fresh feed is to rotate your pastures. By rotating between several pastures, you may maintain your hens healthy and in a dynamic environment by avoiding overgrazing and enabling the pasture to recuperate. For maximum results, change pastures every two or three days, or even every day. Your chickens may enjoy a safe, diverse, and fascinating environment with regular pasture rotation that also assures their safety and availability of a range of natural food sources.

70

Include Technology

Using technology to your advantage may enhance the way you keep an eye on and protect your flock of free-range birds. Motion-activated cameras can inform you of any dangers so you can take prompt action to safeguard your chickens. You may care for your flock more swiftly and effectively while assuring that they acquire the correct nutrition by putting automated feeders and waterers for your hens.By introducing technology into your free-range management, you can keep an eye on your hens more effectively and make sure they're safe,and healthy. Over the years, chicken keeping has evolved tremendously, making it easy to become a new enthusiast.There are various benefits to keeping free-range chickens, such as richer eggs, happier flocks, and healthier birds. You can establish the best environment for your free-ranging flock to thrive by correctly setting up your yard, putting basic management strategies into effect, picking the appropriate breed, and striking a balance between safety and freedom. Enjoy the advantages and joys of free-ranging while taking care of your favorite chickens' safety and well-being.

What people need to know about free ranging chickens

Does Free-Ranging Produce Better Chicks?

Free-range chickens do profit from it! It helps youngsters to receive the psychological and physical rewards of an enriched environment. Because they may seek food in their natural settings, healthier hens produce eggs with a larger diversity of nutrients and flavors.Free-ranging chickens are a substantially better alternative for satisfied and healthy birds because they have access to sunshine and fresh air.

Which Chicken Is Ideal for Free-Ranging?

The appropriate hens for you will depend on what type of free-range lifestyle you require. But among the most well-liked breeds for foraging and wandering freely were **Australorps, Americaunas, Buff Orpingtons, and golden-laced Wyandottes.** These breeds are all suitable for free-ranging as they have calm personalities, make outstanding foragers, and adapt well to diverse climates.

Does a chicken require an acre to be able to wander freely?

For maximum health and laying, free-range chickens require a huge amount of territory. According to a study, 10,000 chickens require roughly 70 acres to wander about in safety. The best ratio for free-ranging chickens is one acre for 500 birds, according to producers. This maintains the surroundings health and offers the birds access to fresh grass and other food sources.

What Drawbacks Come With Having Free-Range Chickens?

Potential exposure to pollution, wild birds and their diseases, predators, and extreme weather are some of the negatives of raising hens in a free-range habitat. In addition, the greater space equals higher housing, food, and animal care expenditures.Small farmers may find this to be an unbearable burden.

Will The Chickens Be Allowed To Stay In The Yard?

Yes, you may allow free-range hens in the yard. If they have had time to adjust to their new environment, they will typically stay near their coop or run for shelter. They should be delighted to walk your yard and cluck along if given the correct care.

Errors to Steer Clear of

It is fairly unusual for a flock of backyard chickens, ducks, or other poultry to suffer nutritional inadequacies. The following five basic blunders individuals commonly make when feeding backyard chickens and other animals are of great nutritional relevance.

1. Not Enough Water

The most vital thing to bear in mind when deciding what to feed chickens is that they require water, and not receiving enough of it may be disastrous. However, until a problem happens, most of us don't give water supply and quality any concern. There are various reasons for deprivation. When the temperature warms, your backyard chickens require extra water, but if you supply the same amount, some of the birds cannot obtain enough. Your birds may not drink the water even if there is plenty of supply if it is too warm. This problem is remedied by bringing out more drinkware, keeping it in the shade, and periodically supplying cool, fresh water.

When the water supply freezes in the winter, it may also result in water shortages. Farm stores and Internet

74

livestock providers sell a range of water-warming devices to meet this problem. Providing your birds with warm, not boiling, water at least twice a day is an extra help. Water deprivation may come from unappealing water, which deters drinking. Providing your backyard chickens with water that you would drink is the best line of action.

2. Not using the Right Ratio of diet

Using a diet that is not suitable for the species, growth stage, or production level of the flock is one of the most common errors made when feeding poultry. What, for example, do ducks eat? What food do chickens consume? Ducks have different nutritional needs than chickens. Any species' chicks have distinct needs from laying hens, which have different demands from a breeding flock. If you purchase ready-mixed feed from the farm store, giving a correct ration is straightforward because most manufacturers offer vital information on the bag or the label. If you opt to combine your diets, you will have to perform a lot of research on the nutritional needs of your other birds, particularly hens, at every stage of their lives.

3. Stale or obsolete ration

A ration starts to lose nutritious value as soon as it is mixed owing to oxidation and other aging processes. Overly chilled feed loses nutrients, gets stale, and loses its taste. In a hot storage room, the process accelerates. Any prepared feed has to be eaten four weeks or fewer after it is ground. Make careful to buy exactly what you can eat in a few weeks, as the farm store may require a week or two for delivery and storage. You may prolong the storage duration in colder months, as I regularly do when winter storms threaten to cut down our distant roads. Feed rotting is slowed down by keeping it chilled and in a closed container. It's useful to know that a vitamin premix has a maximum shelf-life of roughly six months if you mix your rations. For that reason, purchasing a premix in bulk is not a cost-effective alternative for a small flock of backyard chickens. Either purchase premix in tiny enough quantities to serve out before the six-month mark, or make arrangements to share with other chicken caregivers who share your ideas.

4. Overindulgence in Supplements
Overfeeding poultry supplements, such as electrolytes or vitamin/mineral supplements, may result in a serious

76

nutritional imbalance. certain vitamins function in conjunction with one another or affect how specific minerals are utilized. For certain minerals to work correctly, other minerals must be present. Conversely, too much of certain vitamins may interact badly with minerals or even be hazardous on its own, and too much of some minerals might impede the absorption of other minerals.Therefore, the needless use of commercial vitamin and mineral supplements or electrolytes may have the opposite effect of what is intended—that of keeping backyard chickens healthy. Give electrolytes sparingly to healthy poultry. Electrolytes are one form of supplement that should never be administered for longer than ten days (unless advised by a veterinarian). Just before hatching season, electrolytes and vitamin/mineral supplements may aid in boosting the nutrition level in a breeding flock, especially if the birds don't have access to fresh forage. Supplements given to hens several days previous to and during a show can help lessen stress levels. Nevertheless, avoid taking any supplements during a performance as the flavor can make a bird that is unfamiliar with its surroundings stop eating or drinking, which would make it more agitated.The best solution to avoid vitamin and mineral excesses or deficits if you design your rations is to use a premix (like Fertrell Nutri-Balancer) that has been properly manufactured. Premixes are supplied in formulae for conventional and organic poultry feed. Administering too much may be just as damaging as using too little, so be sure to carefully follow the label's directions to avoid overdosing your backyard chickens.

5. Overindulgence in Treats

Everybody appreciates watching their backyard chickens leap into action when we feed them food. However, overindulging in sweets is similar to "killing with kindness."

Feeding too much scratch grain is the most frequent overkill. It's OK to give your backyard chickens some scratch every morning to retain their friendliness. It's alright to feed them a little in the evening to bring them into their coop so you can close them up for the night. Giving your birds a little scratch before bed will help them keep warm on the roost throughout the night in cold weather. However, giving a backyard flock with scratch grains as their major nutritional source does not result in a healthy diet. Similarly, backyard chickens may benefit tremendously from most **kitchen trash.** The leftovers add variation to the **birds' diet**, they are a wonderful source of nutrients, and the birds enjoy fresh fruit. Thus, you may feed your birds kitchen scraps, just as with scratch, but only in small quantities.

Vitamin Supplements for Roosters

Enhancing the Health and Vitality of Poultry

Like any other animal, roosters require specific care to preserve their health and energy at their optimum. Giving these animals food supplements that are suitable for their diet is one of the most common techniques for providing for their requirements. These commodities, which includes,vitamins, minerals, and amino acids.Gallo Biotic Multivitamin for Roosters and Hens is one of the vitamin supplements particularly formulated for roosters. It includes several helpful substances, including vitamins, minerals, amino acids, glucosamine, and chondroitin. It is also fortified with liver, maca, calcium, and biotin to increase physical performance.

Nutritious Supplements For Hens

Amino acids, minerals, and vitamins to boost birds' health and vigor. Animals like roosters require a balanced diet to be robust and healthy. These animals' meals need to be supplemented with vitamins, minerals, and amino acids, among other nutrients. Supplemental vitamin D gives the nutrients essential for the roosters' bodies to perform correctly.

For instance, vitamin E is required for the immune system, but vitamin A is vital for growth and bone health. Minerals like zinc for plumage growth and calcium and phosphorus for bone health are also necessary.The building blocks of proteins, which are essential for the synthesis and maintenance of muscles and other biological structures, are amino acids.

Vitamins that roosters require

Particular vitamin supplements for roosters supply the proper proportions of vitamins, minerals, and amino acids to promote healthy bodily function. They comprise vitamins, particularly **B-12**, which are needed for growth and reproduction. Additionally, vitamin K is present, which is important for blood coagulation. It also includes critical components like iron and copper for the formation of hemoglobin, as well as calcium and phosphorus, which are required for strong bones. It needs these minerals to ensure optimal blood oxygenation.

Appropriate feed for chickens

Keeping roosters healthy and vigorous needs sufficient food. These animals require a balanced diet that is both diversified and full of the nutrients they need to operate effectively.

Corn, soybeans, wheat, and barley are a few of the items that roosters may devour. These nutrients give roosters the protein, carbohydrates, and fats they require to function effectively. Incorporating fruits and vegetables may also help youngsters gain the vitamins and minerals their bodies require.

Products for wound healing and skin care

Maintaining the health of hens needs both feather care and wound healing. For this, specific commodities like vitamins, lotions, ointments, bandages, and healing sprays could be employed. These commodities help in wound healing and infection prevention.It is necessary to note that some products may be detrimental to poultry if not used correctly, therefore before using any product, it is vital to thoroughly read the usage recommendations and contraindications. Furthermore, it is suggested to continually clean and disinfect the facilities and cleaning supplies.

Essential medical devices and supplies for sickness prevention

To prevent and treat sickness, chickens must obtain medical care. Syringes, needles, thermometers, sample materials, and materials for microscopic inspection are examples of necessary medical equipment.Furthermore, there are products made expressly to prevent disease, like as dewormers, vitamins, and vaccines. To identify the right administration and use of these commodities, it is required to speak with a veterinarian who specializes in poultry.

vitamin supplements to increase health and productivity

Supplementing with vitamins is a simple and effective technique to boost the health and production of hens. Amino acids, minerals, and vitamins are included in these supplements. These nutrients increase the quality of the plumage, boost the immune system, preserve bone health, and stimulate egg production. **The Vita-Forte Multivitamin** for Fowl is one of the specialized vitamin supplements that is specified. Additionally, vitamin supplementation may promote blood oxygenation, raise attentiveness, and drive development and hunger in all classes of poultry. It's vital to follow to the product's usage directions and contraindications, nonetheless.

Vitamin Usage And Sales For Hens And Chicks

How to correctly supply vitamins and minerals to roosters It's vital to employ vitamins for poultry according to the dosage and distribution plan specified by the producer. To ensure optimum absorption, vitamin supplements should be taken with water or chicken feed according to the manufacturer's recommendations. To get the most out of vitamins for poultry, it's also advisable to mix them with a healthy, well-balanced diet.

VITA-FORTE Multivitamin for Roosters

For birds deficient in minerals, vitamins, or amino acids, or for those who wish to keep nutritional balance,

VITA-FORTE multivitamin for roosters is a complete oral solution. helpful for breeders, training birds, and roosters recovering from illness.It promotes appetite, maintains nutritional balance, is ideal for developing birds, and is filled with iron, calcium, and liver extract.

Where can I acquire vitamins and other things for caring for chickens?

You may acquire supplements and other things for caring for poultry online at **Vitopharma** USA. This site gives a large diversity of items concerning the health and maintenance of hens, ranging from vitamins and minerals to cosmetics and treatments for wound healing.When making a purchase, it's vital to check the available payment choices and the delivery conditions. To further assure product quality, it is advisable to examine the products' expiration date before purchase. To answer the expectations of farmers and livestock owners, our online store offers a vast assortment of goods. It's vital to check the company's shipping procedures and available payment alternatives before making a purchase. Before placing a purchase, be careful to read all of the information because rules may change depending on the location of the transaction.

Common Inquiries Regarding Vitamins For Cocks

What is the impact of rooster vitamins on hen egg production?

A mixture of vitamins and minerals called chicken vitamins serves to boost the health and vigor of chickens. The hens' overall health and well-being may be boosted, which may raise egg yield.

What benefits might rooster vitamins supply to the full nervous system?

Minerals present in rooster vitamins assist the well-being of poultry's neurological systems. Iron and vitamin B12, which are regarded to be vital to avoiding nervous system damage in chickens, is contained in the specifically created supplement Supertonic Full B12. This vitamin also prevents anemia and improves the appetite of the bird.

In what ways could vitamins aid cocks in terms of growth, blood oxygenation, and alertness?

The minerals and vitamins contained in Vitopharma products are vital for poultry blood oxygenation, development, and attentiveness. One of the finest strategies to improve blood oxygenation and muscle mass growth in roosters is to utilize a red rooster stimulant. Iron and oxygen are better absorbed when vitamin B12 is available, which promotes blood oxygenation and considerably boosts the attention of the birds.

When administering vitamins to hens, what safety procedures should be followed?

It's vital to stick to the suggested doses. Certain vitamins and minerals, such as vitamin A, may have detrimental effects on a bird's health when eaten in excess. After feeding cockerel vitamins, if any unexpected symptoms occur, it is advisable to consult a veterinarian immediately away for additional instruction and, if required, treatment.

Chapter 3

How to Naturally Inspire Chickens to Lay Eggs

I started reading up on techniques to naturally promote egg production in chickens after our hens stopped producing for several months. I wanted to find out what I may feed my chickens to encourage them to produce more eggs as I loathe eating store-bought chicken eggs. I uncovered some practical tips on how to persuade our chickens to produce more eggs naturally. And what do you know? It was successful! We were able to restart having enough eggs to feed our family and quit eating the pale yolk store-bought eggs after applying a couple of these tactics since our chickens' egg production improved. Hooray! A blue porcelain dish loaded with colorful eggs—green, blue, white, and brown—is put on a wooden board. The cause for the chickens' discontinuation of egg-laying is explored, along with many natural approaches to enhance egg production. **There are various reasons why chickens may cease laying eggs entirely or slow down;**

Molting

The fall is the traditional season for chicken molting. When a chicken molts, its feathers fall off and grow back in. During their molting process, hens stop to lay eggs.

Less light

Chickens slow down or halt laying eggs during the winter as the days are shorter and there is less daylight. When they have 16 hours of light and 8 hours of darkness, hens lay the most eggs.

Insufficient calcium

Because the shell is comprised of calcium, chickens require it to produce eggs.

Low protein

When hens are permitted to roam outdoors pecking and scratching for bugs, they naturally produce more eggs, which increases the quantity of protein in their diet. They are unable to regularly seek for bugs in the winter as the ground is frozen and covered with snow.

Stress

When under stress, chickens will stop to produce eggs. Illness, injury, or the risk of a predator are examples of stress.

Age

Older hens tend to produce fewer eggs, so you may want to find out how old your flock is. Less than two years old is considered the best age for egg production, and after that, they tend to slow down.

How to Naturally Raise the Production of Chicken Eggs

Have you ever wondered what to feed your chickens to naturally maximize the quantity of eggs they produce? After giving them a vacation over the winter, we take a few natural techniques to encourage our hens to begin producing more eggs.

Boost Your Protein

We supplement our hens' winter meals with extra protein in the form of sunflower seeds and scratch. We also fed them mealworms one year when we got one of these enormous bags of dried mealworms from someone. Although feeding the chickens dry mealworms appeared a little painful, they enjoyed them! We observed that giving the chickens more protein twice a day benefited. Give them additional protein in the morning and again in the afternoon.Providing the chickens with supplemental protein twice a day benefits in improving egg production. An egg takes a lot of protein and is formed over an entire day. By supplying your chickens with additional protein to begin the egg-laying cycle for the following day, you may hasten the egg-laying process by feeding them later in the day after they lay an egg.

Providing Calcium

Because calcium is the building block of eggshells, chickens require it to produce eggs. Increasing the quantity of calcium in their diet boosts egg production and fortifies the eggshells.

Each chicken coop has a cup that we glued to it, which is supposed to collect crushed oysters or egg shells that have been cleaned and left to air dry.

Add More Lighting

To enhance egg production in the winter, some people offer extra light to the chicken coop. Because there is a danger of fire, you must take great care when using any sort of light or heat in the coop. These extra light sources are now produced by various manufacturers expressly for chicken coops that promote safety. When we sold a lot of eggs, we used to supplement the light in our chicken coop, but we no longer do so. Given how hard our hens work to generate eggs for us throughout the year, we decided to provide them a much-needed rest from egg-laying in the winter.

Whatever fits you and your hens in the backyard is what counts most.

Establish a sanitary and safe henhouse

It's vital to offer hens a clean, safe coop as under stress, they will either stop laying eggs completely or slow it down. Make sure the predators can't access their chicken coop. You may encourage your backyard chickens to produce more eggs by cleaning their coop and providing them access to lots of fresh chicken feed, new bedding, and fresh water.

Raise chickens that have been bred to produce more eggs

Over the years, as we've experimented with producing different sorts of hens, we've discovered that some breeds in our flocks produce more eggs all year round. Our Ameraucanas, Swedish Flower hens, and "Olive Eggers," which we made by mating our Ameraucana hens with our French Cuckoo Marans rooster, are the hardiest egg layers that we have.Compared to all of our other chicken breeds, our Black Copper Marans hens produce the fewest eggs year-round, but they do produce lovely dark chocolate brown eggs.A few types of chicken that are designed for greater egg production than others are Rhode Island Red, Leghorn, Ameraucana, Australorps Brahma, and hybrids like Red Star and Black Star.There may be a problem with a younger hen if she doesn't lay for more than a few months despite having access to calcium, increased protein, lots of light, and a safe and clean coop. Maybe you should take her to the _vet_.

How to increase the quality of newly deposited eggs

You own a flock of chickens that produce stunning, fresh eggs. However, everything you do influences how long your fresh eggs will remain fresh, from where you store them to how you clean a dirty egg! Eggs are safe to eat and fresh when handled and kept correctly. Proper preservation of eggs before incubation is critical for sustaining fertility when hatching them. However, a huge number of chicken caregivers lack essential information. In this section we'll cover everything from washing eggs to keeping them in the fridge. We also go over the best methods to preserve eggs and
answer all of your commonly asked concerns.

Both Clean And Filthy Eggs

A dirty egg is detested by everyone. However, utilizing numerous common cleaning treatments on eggs could make them less safe to ingest!

<u>Is it essential to wash fresh eggs before keeping them?</u>

Washing fresh eggs is not suggested unless they are filthy.The bloom is the term for the thin layer that covers chicken eggs. The bloom inhibits bacteria and other undesirables from getting inside the egg by sealing the porous eggshell. The bloom will be eliminated by washing the egg, enabling pathogens to enter more easily. Even the temperature of the water used to wash the egg could introduce bacteria! So the best line of action is to keep eggs clean. Since the natural bloom will defend the eggs from infection and retain their freshness, clean eggs don't need to be washed.

<u>Before usage, should chicken eggs be carefully cleaned?</u>

Washing chicken eggs before usage is not essential. Rinsing eggs just makes it more probable that bacteria will infiltrate through the shell. Before starting with your cooking, use clean eggs and wash your hands after handling raw eggs or egg shells.

How to wash dirty eggs

Keeping your eggs clean from the outset is the best line of action. Below are some tips on how to achieve this.

But occasionally you get a soiled egg, even in the cleanest coop.

If the egg is unclean

As fast as possible, clean it. Bacteria are more prone to grow and maybe migrate within the egg the longer there is feces on the shell.Filthy eggs are best cleaned by dry scrubbing. To remove any dirt or feces, use a dry paper towel, scrub brush, or dry sponge.Try soaking a paper towel with warm water to get at really recalcitrant feces.Water should only be utilized as a last choice when it comes to eggs. If you must wash an egg, fill a sink or other container with warm water.Never immerse eggs in water that is running, such as from a faucet. Additionally, always use warm water to wash or rinse eggs rather than cold. Eggs become less safe to eat when bacteria are driven through the eggshell by either cold or running water. Even after washing a dirty egg, you need to:

Don't mix it with the clean eggs.

Put it in the refrigerator.(Not raw eggs)

Use it straight away or as soon as you can

Once you have touched the egg, wash your hands. Before ingesting, ensure sure the egg is properly cooked.

Eggs from commercial farms are typically processed for industrial use if they contain more dirt or feces than can be eradicated by dry cleaning. Excessively dirty eggs are judged unsafe and are thrown away.

What to do so your chickens produce the cleanest eggs

Numerous circumstances may result in unclean eggs.Usually, a hen pooping in the nesting box is the cause for a huge piece of feces sticking to an egg. Chickens commonly leave uneven feces imprints on eggs because they drag a tiny quantity of excrement into the nesting box on their feet. An unclean bum is frequently what generates poo streaks on an egg.You may take the following procedures to ensure that your chickens produce the purest eggs.

➢ **Regularly harvest eggs**

➢ Regularly clean nesting boxes and restock the nesting material.

➢ To avoid chickens from dragging excrement into the nesting boxes with their feet, keep the coop dry and tidy.

➢ Never allow hens to sleep or hide in the nesting boxes.

➢ Don't overfeed your chickens with leftovers; instead, offer them a balanced diet of full-layer feed.

➢ Regularly examine your chickens for parasites.

➢ Ensure that your flock has a suitable quantity of warm nesting boxes.

➢ Don't just place broody chicks in the nesting boxes—take care of them!

➢ Set aside the setting hens from the flock as a whole.

➢ Keep the coop clear of insects.

- ➢ Remove a dirty chicken from the herd and contact a veterinarian if it has one.
- ➢ Nesting boxes that are transportable
- ➢ When an egg is laid in a roll-away nesting box, it instantly rolls into a covered compartment. This keeps eggs cleaner and minimizes both unintended breakage and egg chomping!
- ➢ In particular, roll-away nesting boxes work nicely if you aren't home all day gathering eggs.

Storing eggs

It's a frequent myth that eggs must be stored in the refrigerator. However, there are a few benefits to refrigerating eggs. We address all of your concerns regarding egg storage and show you how to increase the shelf life of your eggs!

Do eggs have to be refrigerated?

Refrigerated eggs keep fresher for longer. However, *fresh eggs don't require refrigeration.* It is recommended to refrigerate eggs to keep them cool during the heat. However, we merely bench them for most of the year. It is advisable to refrigerate eggs if you won't be using them for roughly a week. After gathering, store them in the refrigerator as opposed to letting them hang out on the bench for a few days.

Purchased eggs are an exception to this rule. If you got eggs from a store where they were refrigerated, you must likewise do the same at home.

What's the best temperature for egg storage?

When kept, eggs require a consistent temperature. Condensation develops on the eggshell when it heats up after being cold, which may lead to contamination and mold development. Additionally damaging are high temperatures and low humidity levels, which promote evaporation and cause eggs to deteriorate quickly. Within a sweltering, dry climate, eggs may lose as much as 0.1 grams of water every day! When keeping eggs, 2 degrees Celsius and 80% relative humidity are the optimum

100

conditions. It may be perfect to keep eggs in your root cellar if you have one.

In the carton, in which direction should the eggs move up?

Despite plain logic, eggs should never be stored with their pointy edges pointing downwards. Nearly every image of an egg you see is the exact opposite of this! The egg's blunt end, which houses the air chamber, should be facing up to help limit evaporation since it is the strongest point.

Where to store eggs?

Eggs may be stored on the kitchen bench. We use this helpful wire rack to assist us in keeping track of which eggs should be used first while keeping eggs on the bench. However, if you wish to store eggs for a longer period, they need to be kept in a cool, stable environment. Although it's good, high humidity has less of an influence on egg freshness than temperature. The optimal spot to store eggs in most families is in the **<u>refrigerator</u>**, although **<u>pantries and cellars</u>** might serve just as well. Eggs kept in the refrigerator have a shelf life that is more than quadrupled! For a more stable temperature, store eggs on a **shelf** rather than in the refrigerator door. It is crucial to always keep eggs in the refrigerator or basement in hot or dry places. The same is true in locations with big temperature fluctuations, such as the summer, when the residence is hot during the day and cools down at night when the air conditioner is working.

The dos and don'ts of egg storage containers

- Select sanitary materials.
- Eggs should be kept in clean containers to minimize contamination as eggshells are porous. For this reason, reusing cardboard egg cartons is not suggested.

Because they are easy to clean, plastic egg cartons and ceramic or plastic egg holders are recommended. Because it is so clean, even this gaudy blue glass number is a terrific deal!

- Select a dry, safe, and well-ventilated container. A secure storage container is crucial since eggs may acquire tastes and scents from the surrounding environment. However, eggs require air to circulate, so avoid keeping them in a sealed container such as Tupperware.The optimum egg containers do not hold moisture and have sufficient ventilation.

- Utilize a plastic egg tray or carton inside a refrigerator compartment or an open plastic box.

- **Do or don't use cardboard egg cartons?**
Your friends and colleagues will shower you with old egg cartons as soon as they find out you own chickens! It is feasible to wash and reuse plastic containers. But although many poultry owners do, cardboard containers shouldn't be.
On the one hand, recycling is excellent and cardboard packaging provides a (more) green alternative to plastic. However, outdated cardboard egg cartons may contain bacteria that might taint eggs or cause sickness to your flock. Furthermore, cardboard retains moisture and scents,

making it a poor option for egg preservation after the initial use!

Thankfully, cardboard egg cartons have a lot of other ecologically good uses, so you may accept your friends' egg cartons gracefully without feeling awful about putting them away!

Handling of eggs

Even with clean eggs, always wash your hands after handling them.The shell may contain salmonella and other hazardous pathogens, however, this is unlikely. This is especially true when there is poo on the eggs. Washing your hands protects you and your food, but the bloom will shield the egg.

ADVICE TO INCREASE EGG PRODUCTION

One of the staples of the human diet and a driving factor behind the local food movement is the humble egg. The eco-agriculture movement would not consider the practices utilized by current industrial farms to boost egg production rates to be normal or humane. However, both commercial and homestead egg producers who care about the environment may make efforts to safely boost laying rates. **Breeding, food, and the comfort and well-being of the birds** are the three key aspects that determine egg production.

BROWN EGGS

Before talking about these three characteristics, let's discuss a little bit about the brown egg, which—rightly or wrongly—has come to symbolize the cornerstone of the natural food movement. The tint of brown eggs may vary widely, ranging from a very light tan to terracotta and deep, chocolate brown. Naturally, they are precisely the same insides as white eggs. Raising procedures of the chickens are practically the sole element impacting flavor and nutritional value. The majority of breeds and crosses that produce brown eggs were developed to be either meat birds or multifunctional birds. A significant lot of work was put into expanding the number of brown egg-laying breeds at the beginning of the last century. However, most chicken breeding since World War II has depended on hybridization to generate an industrialized bird for the laying plant and colony house. Although these chickens resemble the part, they aren't the same as they once were—they may be even flying by on performance data from those earlier generations of birds.

It was never meant for the larger-framed, heavier-bodied brown egg hens to compete as layers against white egg breeds like the Leghorn. They require greater housing and nest space to be in excellent health, have larger frames, grow more slowly, and consume more feed as they age. Additionally, they lay fewer eggs per hen. Due to their greater size, they may exhibit a degree of resilience that birds with higher metabolisms—some of which are even hybrids with brown eggs—may not have. After only one laying season, many flocks of brown egg-laying hybrids are turned over due to their fast burnout factor. Unfortunately, they generally have very little salvage value owing to their microscopic size and the wear and tear of intensive manufacture. Purebred brown egg layers with stronger laying lines are frequently those from basic breeds **like Rhode Island White, Australorp, and White Plymouth Rock.**

IMPROVING EGG QUANTITY THROUGH BREEDING

Breeders may progressively boost the output of these classic brown egg-laying breeds' flocks. Replacing pullet chicks for reasonably large laying flocks may be generated by a small flock of well-bred females. Since the initial farm offers the most critical input—the seed stock—such an attempt is viable. It all boils down to finding out which females are the most productive and which of their male descendants to employ to establish a line that performs on the home farm.

Increasing egg yield via reproduction

The first step in enhancing egg yield is good breeding. For many years, the Hogan technique has been taught as a way of analyzing the capacity of young stock to layer and the continued success of generating chicks. Although the process is fairly easy to teach, it involves a lot of work because each bird must be inspected by hand. Work should be done on the laying flock regularly to cull out underachievers and ill or damaged birds. This ensures that critical feedstuffs only reach the birds that are utilizing them to their maximum and most lucrative benefit. Just like the producer of outstanding Hereford cattle or blooded horses, the egg producer must take breed selection and performance development carefully. The answers to your queries about laying performance are there in front of you whether your flock is dominated by roosters, is made up of hens that are passed the second year of laying, or has at least one of each breed listed in the large hatchery catalog, or if foxes and raccoons are the only ones who do the culling! A competent egg producer needs to become a competent poultry breeder to become sustainable and establish predictable performance.

A word of caution when ordering chicks

Well-bred pullet chicks may now cost $7 or more each. Good foundation stock is not affordable. Lately, I've observed several adult breeding bird trios (one male and two women) that cost at least $75 to $100. For hens? Yes, but, keep in mind that this is still considerably less expensive than what even the most ordinary feeder calf would generally cost.

109

FEED TO IMPROVE EGG PRODUCTION

At every stage of development, chickens require a well-thought-out nutrition plan as high-quality feed is the source of their eggs.

Increasing egg yield via feeding

When it comes to layer feeding, never skimp.

Every day, hens consume a small quantity of feed, hence their diets need to be nutrient-dense and homogeneous in shape. A hen in lay will take 4 to 8 ounces of feed per day, depending on her size and breed. So, those trying to cut feed costs should start with birds that will deliver eggs in a manner that is highly feed-efficient. To establish the exact amount of feed necessary to yield a dozen eggs, the producer is required to keep goods records.Recently, there has been debate over some pretty strange combinations for poultry feeds. These regimens might be viable for a restricted number of niche markets, where the customers can pay a premium that balances the increased expenditures of these feedstuffs. Some may be fairly costly to make. It may be essential to acquire specially made rations in quantities as small as one to three tons if components are not widely available. The old maxim suggests that a producer must produce at least 100 tons of feed per year to fund the on-farm equipment necessary for processing.

The present livestock era has been marked by constant breakthroughs in the quality of poultry feeds. Nutritional advancements were frequently applied initially to laying hen and newborn chick feeding. Currently, a number of the top feed producers offer kelp and fish meal-fortified diets, feeds with higher omega-3 content, and blends of chicken feed that are fully vegetable-based.

Here are many crucial aspects of the poultry diet

Start with a quality **chick starter** that you purchase in small numbers to ensure the supply is fresh. Today's starter/grower meals are meant to be served to young pullets until they lay their first eggs. These better meals fulfill the twin objective of expanding the egg tract and the frame. The young females should be gradually transferred to a quality laying ratio as the first eggs mature. Some farmers are reverting to the traditional approach of feeding their chicks numerous times a day using hard-boiled eggs that have been severely chopped.

111

Offer no more finely chopped egg than the chicks can take in roughly 20 minutes at feeding to guarantee wholesomeness. This is especially good for stressed-out chicks that had terrible transportation. They should shortly be moved to a full starting ratio that is provided to them at their option.Using rations that are meant to be small or micro pellets may help cut down on wasted feed. Pelleted feed makes it simpler for birds to grab food pieces that they flip out of the feeder.One technique to guarantee that rations stay fresh and to spread out costs over a year is to buy feedstuffs at roughly two-week intervals, if practicable.All feedstuffs should be **kept dry and protected from rodents once they are at home**. Most feed varieties will keep no more than 300 pounds in a 55-gallon barrel. These days, the majority of complete chicken diets are fortified with the required minerals and grit. Oyster shells were once regularly offered as presents, however, they were occasionally sold in forms that were too huge for hens to ingest.Grit is given by numerous older workers who merely dump stream sand into low-sided wooden bins that the birds can access. A clean grit product that is cheaply priced is cherry granite grit which is the proper size. Nowadays, many feeding regimens do not require scratch grain. It is favored by birds over complete meals, which boost egg-laying performance, but excessive grain intake may lower egg output. To entice the birds back inside the coop, it would be good to offer them as much grain as they can consume in roughly twenty minutes, as well as to hand it on to them at the end of the evening. This adds to the birds' already hot vigor as they move into a frigid night.Older chickens or hens kept for other reasons don't produce many eggs, but far too many farms continue to supply costly feed to birds that don't need it. Cost-cutting should never be done on feedstuffs or seed stock. Providing the birds with clean, fresh water to drink and food items is the first step towards enhancing egg production. Producers continue to learn the skills required to pick which birds to replace, when to replace them, and how to develop better replacements.

Chapter 4

Learn How To Identify And Steer Clear Of The Most Common Illnesses That Affect Chickens

One of the most important aspects of raising hens is being able to recognize and prevent the most frequent diseases that might affect your flock of hens. By the end of this chapter, you will know the necessary to identify the most common diseases that affect hens and to give preventive measures. Poultry, like the progeny of your cat and dog, is susceptible to bacterial, viral, and parasitic infections. These ailments may affect poultry too. Because hens are considered to be food animals, there are not many therapeutic options that are currently accessible. Given the constraints that are placed on the treatment of infectious illnesses in hens, it is of the utmost importance to emphasize sickness prevention. Through the implementation of a disease prevention plan, you have the potential to reduce the risk of the sickness being passed on to your birds, as well as the degree of the illness experienced by them.

Characteristic signs and symptoms of illness in chickens

A prominent sign that your flock is unwell is the presence of symptoms that are associated with respiratory or non-respiratory infections.Because the symptoms of respiratory disorders sometimes match those of other diseases, recognizing one without undergoing a necropsy may be challenging.

Signs of Respiratory Infection

peculiar breathing sounds referred to known as "rales"

Shaking of the head

dirty wings (caused by birds plucking at their feathers)

Coughing

sneezing

the slow rate of growth

purple-blue discoloration on the face

discharge from the eyes

edema in the face or wrists

"Conjunctivitis" is the word for inflammation of the eye conjunctiva.

discharge of the nose

Signs of a less common respiratory infection

swelling joints

Warts and scabs

Twisting of the neck and head, or "torticollis"

Spots of red or white on the legs and comb

Immobility

Laziness

Diarrhea: It could look green and watery.

Signs of Non-Respiratory Infections

Strong smell

difficulties breathing

Reduced consumption of water

tangled plumes

Lack of water

Feathers twisted or broken,   described as **"helicopter wings"**

Reduced intake of food

A rise in deaths

hindered weight loss or growth

Reduced production of eggs

Debilitation

shell-free eggs

Blue and purple discoloration on the face

Eggs with thin shells

pale comb

Wet poop

tiny comb

maybe green diarrhea

Infected wattles

Deficiency

oral discharge

The twisting of the neck and head is termed "torticollis."

Lack of vision

Immobility

expanded stomach

Shivers

navel infection

Laziness

swelling joints

Footpad edema

Which Diseases Affect Chickens the Most Frequently?

As usual, visit your poultry veterinarian to decide the best course of action for treating infections in hens. It's vital to learn the most prevalent illnesses that afflict hens and how to prevent them.

Marek's Syndrome

Marek's disease is a highly transmissible viral virus that usually affects chicks but may infrequently afflict older poultry. It generally affects young chickens with clinical indications showing between 6 and 30 weeks of age.

Signs and symptoms of Marek's syndrome

Tumors resulting in paralysis and death have been related to Marek's Disease in chickens. To diagnose Marek's Disease, vets must undertake a necropsy. If your baby bird is paralyzed and has not undergone vaccines, it is most likely related to Marek's Disease. Marek's disease has no known treatment, although it may be **prevented** by Vaccinating your birds: immunizations need to be delivered **"in ovo," which is on the day of the chicks' hatch or during the egg's incubation.**

As soon as you introduce new birds or chicks to your coop, remove any feather dandruff.

coccidiosis

Chicken coccidiosis, which primarily affects younger chicks but may also strike adults, is caused by an abundance of species of coccidia, protozoan parasites that infect the gastrointestinal system.

Signs and symptoms

In addition to reduced water and feed intake, decreased egg production, weight loss, anemia, poor growth, diarrhea, and even death, these organisms live in your chickens' digestive systems.

Prevention

Stop water from spilling. Coccidia develops in water coupled with chicken feces. Keep the coop's soil as dry as possible, particularly in spots where the birds gather, such as feeding and nesting grounds. To help minimize coccidia infection, use high-quality coop bedding. Fresh bedding absorbs excess moisture.Feeding medicated chick feed,may help eliminate dangerous protozoa in the growing birds' gastrointestinal tracts and enhance their natural defenses.

Bird flu, or avian influenza

Highly Pathogenic Avian Influenza (HPAI) is typically recognized as the most deadly infection to strike poultry worldwide. Each year, the virus kills a considerable number of farmed birds.

Signs and symptoms

Egg-laying chickens may display malformed or soft eggs and/or a considerable decline in egg output. Other signs of bird flu include dehydration, not eating, rapid death, lethargy, diarrhea, leg or foot hemorrhages, respiratory difficulty, and head inflammation with blue-toned combs.Note that symptoms of avian influenza may also be shared by other infections; contact your poultry vet to confirm a diagnosis of bird flu.

Prevention

All poultry owners must adopt preventative procedures to avert bird flu infection in their flock because there is now no vaccine or therapy for the illness. These precautions include:

not exchanging animals for feed, supplies, or other chickens with other chicken owners.

Cleaning and sanitizing shoes, clothing, and automobiles thoroughly to stop sickness from entering your flock (if you've come into contact with ducks.)

Keeping your flock of birds inside their coop and fence helps keep them away from wild birds.

placing any new birds under a 30-day quarantine and screening them for sickness before introducing them into the flock.

Make sure the housing for your flock is distant from any bodies of water to prevent getting into contact with ducks.

To halt sickness spread, assess your flock regularly and remove any birds displaying symptoms straight away.

working together with state and federal veterinarians, county agricultural extension organizations, local institutions, and your veterinarian.

Newcastle disease with virus (vND)

Even while vND strains are infrequent in North America, they may cause major health concerns to backyard and domestic birds when they do emerge.

SIGNS AND SYMPTOMS

edema around the eyes and neck

diarrhea with a watery, green tinge

shaking, absolute stiffness, spinning, losing wings, and a decline in activity

breathing difficulties, coughing, sneezing, and nasal discharge

A rise in flock mortality and/or unexpected fatalities

Prevention

As a complement to established measures in management and biosecurity, vND immunizations are indicated for backyard birds, especially those that dwell near afflicted flocks.Many feed stores carry vaccinations against Newcastle disease, such as the B1 and LaSota vaccines. vND is spread by infected bird excretions, feces, and aerosols; as a consequence, the virus may be discovered in shoes, tools, clothing, the environment, contaminated feed, water, and equipment.The sickness is exceedingly contagious and has no recognized treatment.

Salmonella

Chickens, humans, and other animal species may all develop salmonella, a common bacterial ailment that is transmitted by contact with polluted surroundings, tainted eggs or animals, or contaminated feed.

SIGNS AND SYMPTOMS

The sickness is **seldom** observed in adults even if they are positive for the germs; symptoms of salmonella in chickens may vary and include: Young birds up to two weeks are the ones most prone to getting the disease and dying. In most situations, the birds do not display symptoms, and productivity is not impaired.

insufficient growth

appetite drop

diarrhea with a lot of water

fragility.

Infected eggs may cause decreased hatching rates and higher embryo death, as well as an increase in mortality in newly fledged chicks.

Prevention

It's vital to wash your hands after touching your chickens since they may have salmonella, which may infect humans.

When the flock starts producing, don't introduce any more birds.

Clean and disinfect in between batches of birds.

All feed should be heat-treated to get rid of salmonella bacteria.

Before entering the chicken coop or enclosure, undertake severe biosecurity requirements such as hand cleansing and changing into new shoes.

To stop environmental contamination, utilize rodent control techniques, such as a chicken feeder that is impenetrable to mice.

Avoid allowing wild birds to approach your chicks.

The feed should never be left outdoors as this might attract rats that may infect your flock with salmonella.

Mycoplasma

Mycoplasma is a bacterial ailment that is typically detected in backyard chickens; culling infected individuals or testing and culling breeding hens are the only options to remove mycoplasma infections in a flock.

SIGNS AND SYMPTOMS

Mycoplasma infections frequently result in both acute and chronic respiratory diseases, which may present as:

discharge from the nose and eyes

sneezing

spitting

sinus infections.

Among the secondary symptoms are:

fragility

skeletal anomalies

inflamed and unhealthy joints in the hock and foot pad

embryonic demise

insufficient laying

Although you cannot assume that your birds are sick merely because they display the clinical symptoms indicated above, other infectious illnesses that afflict poultry, such as chicken cholera, infectious bronchitis, infectious coryza, and Newcastle Disease, also have similar clinical signs.

Prevention

The prevention of mycoplasma demands severe biosecurity measures like:

utilizing footbaths to disinfect

sports shoes and clothing that are different

having separate equipment

restricting access for individuals.

No birds from your flock should be exchanged with other persons who own poultry flocks, or taken to exhibits, auctions, or flea markets if your flock is mycoplasma positive.

Mycoplasma is exceedingly contagious and commonly affects backyard flocks, but thankfully does not kill many birds. To guarantee that your birds do not become sick, you must adopt proper biosecurity procedures.

How does biosecurity work?

When it comes to flock health, biosecurity is all about taking safeguards to keep people, property, and birds safe from infections and viruses.

The United States Department of Agriculture (USDA) describes biosecurity as a two-pronged defense that consists of:

Plan for Biosecurity in Chickens

A method for chicken biosecurity assists in keeping healthy, disease-free farmed poultry. The following are USDA-recommended best practices for poultry biosecurity:

Report ill birds. Wash your hands before and after handling live birds. Wash your hands with water and soap before and after handling live poultry. Call your state veterinarian, local veterinarian, and/or cooperative extension service office as soon as your birds become sick or die away. For everyone who comes into touch with your flock, supply disinfecting footbaths and/or disposable boot coverings. Make sure the footbath is maintained clean and that any dirt, mud, and/or droppings are removed from shoes/boots before immersing them in the footbath. Change clothing both before entering and after leaving the bird habitats. Visitors should put on protective apparel or disposable headgear, boots, and coveralls before entering the poultry plant.Control visitor interaction. It is ideal to have as little touch as possible between guests and your birds. Anyone who deals with them must observe biosecurity procedures.It is your obligation as a "flock keeper" to make all reasonable efforts to keep your birds healthy, such as feeding them correctly, keeping a clean and dry coop and, vaccinating them against specific illnesses, putting a biosecurity policy in place, and seeing a veterinarian.

First Aid for Chickens

Taking Care of Sick or Injured Chickens

The bulk of us dedicate a lot of time to getting ready for the arrival of our first chickens, but relatively few of us ponder how we would manage significant illnesses, accidents, or end-of-life choices until confronted with them. It may be quite tough to locate a veterinarian who is educated in poultry medicine or any doctor who would treat a chicken; nothing makes a chicken keeper feel more helpless than not understanding how to aid a flock member in need. Being ready to confront the most hard portion of chicken-keeping when the time comes might reduce the strain of an already tough season. Although hens are good at disguising sickness and wounds, if you spend enough time with your flock, you will learn to notice the subtle indicators that anything is wrong.

Typical indicators of a sick hen include:

concealing

Pale wattles or comb

odd feces

weird alignment

lethargic

lack of hunger

the reduced yield of eggs

All of these imply that additional in-depth examination is necessary.

Getting Ready for Accidents and Diseases

Keep first-aid equipment ready so that you can offer emergency medical care in an emergency and maybe save the life of an injured or unwell chicken. The normal chicken keeper can tend to ill and injured chickens, at least until you can get the essential veterinary treatment.

Put Together A First Aid Kit

First-aid materials are best maintained in a plastic container with a lid. The components of your first-aid pack should include:

Antibiotic ointment (free of chemicals that particular birds may find harmful to pain treatment)

single-use gloves

dog nail clippers (for cuticle or beak trimming)

Epsom salt (to soak certain wounds)

eye dropper or syringe (for delivering liquid sustenance, medications, and water by hand)

LED torch

non-stick gauze pads

Infant bird formula in powder form (for hand-feeding)

cutlery

bandages that stick to themselves

Styptic powder (for beaks or nails that bleed)

The super-glue gel is used to mend injured beaks

tweezers

vitamins and electrolytes (to avoid dehydration, shock, and heat stress)

Establish A Recuperation Area

A quiet area within the home, garage, or basement works well for a recuperation place; it should be predator-proof, lined with soft bedding (such as puppy training pads, pine shavings, new bedding, or soft towels), and easily positioned for regular inspection. Determine where a sick or wounded chicken may be temporarily placed, away from the coop and other flock members.It is efficient to employ a folding, portable chicken kennel/cage since it is lightweight, portable, and easy to clean. To ensure dry bedding and clean food and water, keep cage cups for food and chicken nipple drinkers for water ready to hang within the cage.

Find Veterinarians in Your Locality

Contact all the veterinarians in your region to find out whether they treat chickens. If you locate one, book an appointment to meet the veterinarian or staff. A simple introduction may make the difference between getting seen in an emergency or being informed the vet only sees "established patients." Poultry vets are hard to come by. Avian doctors are easy to locate, however, not all avian physicians are trained, skilled, or comfortable treating chickens. State veterinarians, poultry extension agents, and state veterinary diagnostic labs are just a few of the state and federal services that may support backyard chicken keepers in diverse ways. The USDA's Veterinary Services offers a free disease diagnosis consulting service with a veterinarian; contact 866-536-7593 to talk with a USDA vet in your region.

Providing First Aid To Your Herd

Remove The Ill Chicken.

ill and wounded birds require a peaceful, safe location where they can be monitored attentively; remove the ill bird from the flock to protect it from other birds' bullying and pecking and, in the event of a sick bird, to protect the flock as a whole from a potentially contagious disease.

Since hens have poor eyesight at night and are simpler to trap then, try again around dusk. If you are unable to catch an injured or ill chicken during the day, it may be in shock or disorientation. To help keep the wounded chicken quiet and safe for transit, wrap it loosely but securely in a thick towel.

Stop the Bleeding

Wear gloves whenever practical when handling a bleeding or wounded chicken; cover superficial wounds with styptic powder and keep them closed until the bleeding stops; treat the injuries with a clean cloth, gauze, or paper towel; and apply firm, constant pressure until the bleeding stops.

Examine and Clean Wounds

Examine the chicken from head to toe. Feathers may obscure wounds, particularly those from a hawk's talons, therefore washing the bird may help discover damage sooner. Wounds could appear much worse than they are until bleeding is stopped and the area cleaned. I used a syringe filled with freshly mixed Dakin's solution to flush and irrigate very deep or highly filthy wounds after properly washing the wound with water, hydrogen peroxide, or Vetericyn Plus Poultry Care spray. Watch for indicators of infection, like as swelling and redness in the location, and keep the incision clean and dry while the bird recovers. I use Vetericyn wound spray two or three times a day until the bird has recovered, but you may use a triple antibiotic ointment. Since wounded hens should always be kept separate, there's no need to conceal wounds; other more effective, less painful wound-care solutions function without hiding the first symptom of infection: redness. I do not advocate using blue or purple antiseptic sprays or liquids on wounds. The hypothesis that the purple pigment covers wounds from other birds, sparing the wounded bird from being cannibalized, seldom proves to be correct. Getting an antibiotic from

134

the feed-supply store can result in the inappropriate kind of antibiotic being provided, putting your bird at risk for subsequent complications. If antibiotics appear needed, contact the poultry agent of your state's agricultural extension office for help.

Make The Chicken Hydrated

Water is involved in every phase of a chicken's metabolism; dehydration renders recuperation an uphill struggle, if not altogether hopeless. Adding a vitamin/electrolyte supplement to the drinking water for a day or two may aid in healing from heat stress, injury shock, and dehydration. Keeping a sick or wounded chicken hydrated is the next concern while caring for them. This may involve often providing water by spoon or dropper. For a sick or wounded bird, food is initially significantly less necessary than water. Once hydrated, if the bird is not eating on its own, you may provide it baby-bird formula by spoon, dropper, or syringe or it can be tube-fed, also known as gavaged, a method that is beyond the scope of this article and needs specialist training.

Reduce Pain

Since chickens are at the bottom of the food chain, they must behave stoically to avoid calling attention to themselves when they are unwell or wounded. Although they don't exhibit symptoms of discomfort, don't believe that their stoicism implies they aren't in pain—chickens are just as pained as you would be if you had an accident or sickness of a comparable sort.For hens, meloxicam is a regularly used anti-inflammatory; however, a veterinarian must prescribe it along with the dose depending on weight and any advised egg deprivation time. The Mississippi State University Extension Service indicates that an injured chicken may be given an aspirin drinking-water solution for up to three days, providing there are no internal injuries, at a dosage of five aspirin tablets (324 mg) per gallon of water.

Take Care of Internal Wounds

You may suspect infection and/or internal problems if an injured chicken does not recover with care or if its health deteriorates. If a bird needs treatment for internal injuries, only a veterinarian can aid.

Don't Change The Chicken's Nutrition

Offering foods or supplements that the bird would not ordinarily accept may radically change its diet, making it more difficult to detect and evaluate the problem and making an already sick chicken feel much worse. Herbal treatments and other nutritional aids should not be provided to ill hens if they haven't previously been a regular part of their health regimen. Once the chicken's health crisis has passed, focus on boosting its immune system. Generally speaking, a probiotic supplement in the water is a safe approach to providing helpful microorganisms to a sick chicken without introducing additional chemicals that might cause medical suffering.

Avoid Medication at Random

Avoid providing dewormers, antibiotics, garlic, vinegar, molasses, yogurt, or any form of herb to a sick chicken until the underlying issue has been diagnosed. Until the underlying disease has been discovered, any therapy, even natural therapies, may cause damage to the bird. Randomly medicating a sick bird without understanding what the underlying issue is might make their health much worse and hinder the ability to pinpoint the underlying

137

problem. Sometimes the best thing we can do for a chicken beyond being rescued is to assist in ending their misery peacefully. Often, the supportive care indicated in this book is all that we can do at home for a sick chicken, and many times, it's enough to bring them through a medical crisis and back to their flock. When it isn't, and veterinary treatment isn't an option, the worst-case situation is always death.

Developing a Bond with a Poultry Vet

Building a good and trustworthy connection with a qualified poultry veterinarian is crucial to prudent flock management. Unlike other pets, chickens have particular health demands that need the expertise and experience of specialists knowledgeable in avian care. Developing a proactive connection with a poultry veterinarian is crucial to the overall health and lifetime of a home flock. Finding a veterinarian with skill and experience in poultry medicine is a vital first step in this procedure. Although conventional veterinarians may have a rudimentary grasp of animal health, a poultry vet is particularly educated to address the specialized needs and issues that are specific to hens. Using resources like veterinarian directories, referrals from area agricultural extension offices, or recommendations from other poultry owners may assist in discovering a suitable practitioner.

It is essential to get in contact with a poultry veterinarian well in advance of any critical health concerns appearing in the flock. This proactive approach fosters the creation of a connection based on mutual understanding and trust. Present yourself as a responsible chicken keeper eager to develop a long-term relationship for the health of your flock. This initial engagement establishes the framework for future open communication and joint decision-making. Scheduling regular veterinary visits, even in the absence of visible health difficulties, enables the veterinarian to examine the general health of the flock, detect possible issues early on, and make specific advice for feeding, immunization, and disease prevention. This proactive approach to health care is a hallmark of prudent poultry husbandry. Routine check-ups and preventative treatment are vital components of proper poultry management. In the case of an emergency or unforeseen health condition, having a veterinarian who is knowledgeable about the flock's specific features is immensely useful in allowing more effective and targeted diagnosis and treatment. Timely veterinary treatment may also be vital in limiting the consequences of health disorders and assuring the speedy recovery of individuals who are ill.

For the relationship to be successful, communication with the poultry vet must be effective. The vet has to be able to make recommendations based on observations made about the flock's behavior, changes in egg production, and any uncommon illnesses seen. The chicken keeper should also be educated and actively engaged in the care process by asking questions about suggested treatments, drugs, and possible hazards. As the connection between the chicken keeper and the poultry veterinarian grows, frequent reports about the health state of the flock and any alterations to management tactics serve to build a conversation that never stops. This cooperative method fosters a feeling of shared responsibility for the well-being of the hens. Furthermore, staying up to speed on innovations in poultry health and management practices helps the chicken keeper's ability to confer with the veterinarian and make knowledgeable judgments. To sum up, having a connection with a poultry veterinarian is a crucial strategic investment in the health and lifespan of a home flock. It is a proactive and responsible approach to poultry management, backed by frequent veterinarian visits, open communication, and a shared dedication to the welfare of the hens in one's care. This collaboration is important to the framework of sustainable and ethical chicken-keeping.

Chapter 5

How to give winter warmth to your hens

Although chickens are hardy creatures, laying hens will stop producing eggs if they aren't kept warm enough, and weaker flock members can grow sick and find it difficult to recoup in the cold. Although some farmers use heat lamps to keep their coops warm, if they fail or aren't maintained appropriately, they run the threat of sparking fires. These seven methods might aid in ensuring that your birds are kept secure from the cold.

Reduce the number of drafts

Your coop's rate of heat loss may increase owing to wind chill. This is why you need to make sure that any air leaks are covered properly as the nights go shorter. If your coop is brand-new, there shouldn't be many gaps; but, if it's older than five years, it's possible that some of the portions have started to rot and will require repair. Putting a piece of plywood that has been cut to size over the holes is the easiest and least costly technique to repair them. To avoid the temperature from lowering too fast, block any extra gaps in your coop, providing your vent is running effectively (see below).

Maintain proper ventilation in your coop

Large apertures enabling cold air to enter the coop are not a good idea, but you also need to be cautious not to limit ventilation too much as this may create severe difficulties like ammonia accumulation. Make sure you have a proper ventilation system in place to avoid this. Vents must be positioned toward the coop's roof, so the cool air can't reach your birds directly. You can manage the humidity and prevent mold from growing in your birds' bedding by venting out the warm, moisture-laden air and replacing it with cooler, drier air. Your mesh vent should preferably contain an opening gate that you can close or open. In this approach, you'll be able to sufficiently open the coop during the day and shut it up at night when it's colder outdoors or after extremely powerful downpours.

Apply the "Deep Litter Method."

In addition to being a sustainable strategy to manage the litter in your henhouse, the Deep Litter Method may assist in providing insulation for your flock in the winter months. First, merely cover the floor with pine shavings or any other comparable organic substance. You simply need to use a light rake to mix up the bedding and allow your flock's natural movement to take care of the cleaning and replacing of the waste that your chickens make.When prepared properly and replaced with pine shavings periodically, the litter will start to build a compost layer that attracts helpful bacteria and helps them to break down dangerous germs present in the chickens' waste. This is not only a lot easier approach to managing waste and helps insulate your coop over the winter, but it may also help prevent mite and lice infections.**Note:Cedar shavings are harmful to chickens, so avoid using them.**

Utilize sunlight to maintain heat.

Winter may bring shorter days, but you may still harness solar radiation to preserve heat during the day and keep the coop warmer for longer at night. When you apply the Deep Litter Method or have a dark, dirt-floored slab, well-insulated windows may create sun traps.Adding extra "thermal mass" to your coop may help it hold onto heat for prolonged periods. The capacity of a material to collect heat and release it later is measured by its "thermal mass." The more thermal mass your coop has, the more reliably it will radiate heat after the sun sets. Concrete, stone, and compost floors are among the materials that will store more heat during the day and release it at night.

Make sure the chickens have a place to roost

Since hens naturally sleep in bunches and puff up their feathers to keep warm, you must give them a place to relax. Generally speaking, your roosts ought to be erected at least two feet above the ground. They feel protected and are kept off the frigid ground by having access to a roost that is above the floor.As winter approaches, it's vital to ensure that every one of your chickens has a nice spot to roost. Use a torch to check on them in the evening to make sure this is the case. There won't be enough space if one is on the ground, hence the roots will need to be extended. It's advisable to erect roosts at least two feet above the ground.

Build a solarium for them

While it may be tempting to isolate your birds from the cold, allowing them to move around more may be beneficial to your flock. You may expand your coop by erecting a "cold frame" or greenhouse-style expansion, covering it with clear plastic to protect it from bad weather and providing it extra area.Your birds will be sheltered from the wind, rain, and snow while still having plenty of space and fresh air.

Prevent frostbite

Breeds with large combs and wattles may be more prone to frostbite during the coldest months or harsh weather occurrences. You may rub petroleum jelly on their combs and wattles to offer them extra protection and fend off the worst of the cold.Your birds should be pleased and healthy during the winter if you follow these suggestions. No matter how cold it gets for the rest of us, you can make sure that your birds keep warm and continue to produce eggs by taking excellent care of the coop and your flock.

How to maintain your hens' cool throughout the heat

Do summertime temperatures in your area soar? Are you aware that chickens may quickly die from excessive heat? Through the frigid winter months, hens become specialists at maintaining their body temperature. Even in the deepest frigid conditions, they are completed with a brief fluffing up of feathers.Heat is a unique concern. When the temperature rises, they have problems keeping their body's heating system cool.

Instead of sweat glands, fowl depend on their wattles and combs to evacuate heat. And sometimes, it simply isn't adequate. At roughly 24°C (75°F), they are still pretty comfortable, but any greater temperature would make them restless.They require support to maintain a constant body temperature once the dial hits 30°C (86°F).They'll be at significant risk of overheating by the time it gets 32°C (90°F). Heat stress will occur if the temperature is high and they receive no support to remain cool.Severe difficulties, including death, may ensue from heat stress.

When should I take action?

Make plans to aid your chickens in cooling off before the weather gets too hot. You're wasting time if you wait until the heatwave comes.It's vital to recognize the warning indications of disturbed chickens and what you can do to aid them. There is no option but to act: heat exhaustion is the cause of death for chickens.Don't solely believe the weather reports. Keep a watch out for behavioral changes in your flock; if you observe pale combs, open beaks, or wings held far out from the body, it's time to act.

Six simple techniques to keep summertime chickens cool

We'll look at six quick and easy procedures in this portion that you can perform right now to safeguard the safety of your chickens come springtime. Even though the methods are basic and sometimes utilize goods you already possess, you must make sure you have the right equipment accessible.

1. Water

2. Electrolytes

3. Fruit

4. Freeze!

5. Dustpan

6. Cooperation

1. Ensure there is ample supply of cold drinking water for them.

As evident as it may seem, chickens drink four to five times as much over the summer than they would in the cold. It's how they remain cool, primarily.Thus, in warmer weather, make sure your flock always has access to cold, clean water, even if that means changing it multiple times a day.Keep in mind that metal containers may heat up rapidly in the sun if you use them. Ensure they will spend the full day in the shade. Consider modifying the manner you offer your flock water throughout the hot season. Using one or two huge containers is not the most efficient technique for keeping things cool. Present multiple shallower, more compact dishes. Whatever you have that can hold water and is "food safe" should be utilized, such as bowls, basins, and pots. Refrain from using plastic that isn't certified as food-safe as it may discharge contaminants into the water. Don't push your chickens to look for water; they need to maintain their energy. Place the containers in different places throughout the course, particularly in regions where they will stay cool. Beneath a bay tree, When a massive block of ice is put in a shallow dish, it will melt gradually in the heat and give cool water.When it's hot outdoors, avoid using apple cider vinegar. Apples are harmless and may benefit chickens in adapting to the heat. Conversely, there are several advantages to utilizing apple cider vinegar, but heat is not one of them. It accomplishes precisely what you want to prevent: it stimulates the chicken's metabolism, which raises their body temperature.

2. Use electrolytes to preserve health despite the heat

A wise way to rehydrate your hens is to add electrolytes to their drinking water if they look to be hot.Electrolytes are best described as a chicken version of Gatorade, without the flavors and colors. When added to drinking water, they help restore the vitamins and minerals that are lost when chickens are dehydrated. However, it's vital to avoid overdoing it as chickens cannot withstand high salt levels.

3. Juicy fruit is a way to offer your flock some love

Currently, summer fruits are abundantly accessible in supermarkets (and on our trees!), and they're a treat for hens, particularly when they're chilly. Because it has such a high water content, watermelon is among the best. It comprises practically minimal salt, no fat, antioxidants, and multivitamins. Ideal summer cuisine for your party! Purchase a large one, cut it, and freeze it. In the heat, it will slowly melt. In addition to eating the liquid while they devour the meat, hens love the sweet taste. Watermelon helps poultry remain cool during the hot heat. One extremely important food to keep your flock hydrated is watermelon.A summer treat is other juicy foods like pears and strawberries, but watch out for overindulging. It may result in waste! My chickens adore figs; in fact, they have their fig tree!

4. Freeze veggies to have a healthy flock and sail through the summer!

Adding some summer veggies to your hens' diet will also help them keep hydrated in the heat. An even stronger protection against heat stroke is to freeze them and utilize them in a delightful summer salad! Once again, the water is drunk by the chickens as they gnaw away the ice block to access the veggies.It isn't complex science.Grab some watery vegetables; tomatoes work well (no, the tomato vine is poisonous to chickens), cucumbers, sweetcorn (not cracked corn; this will raise rather than lower the temperature of your flock), lettuce (which gets mushy when frozen but still tastes good), shredded carrot, courgettes (US zucchini), etc.

Pour in some water.

Put it to sleep.

When there's a frozen veggie treat on the menu, my chickens want to be the first ones at the table!

5. Encourage summertime hen care with a dust bath

The terrible reality is that rising temperatures encourage mites and lice to chicken coops. Furthermore, at a time when hens are exerting all of their energy to resist the heat, those annoying critters cannot only irritate but even kill them.

Dust bathing is a fantastic approach to aid your flock in remaining cool in the summer and getting rid of mites.Even though I give locations for dust bathing, such as a sand hole for youngsters, wooden blocks, and old tires, my flock frequently constructs its location.Make sure the shade is accessible for your summertime dust showers.

6. Maintain the cooling of your coop!

It's vital to verify that your chicken coop has appropriate ventilation, especially during the heat. The airflow from vents higher than the height of a chicken head should be adequate. Try using a fan to blow some cooler air into your coop if you don't have one and it is powered. Something like this would be amazing.Duringg the day, leave the coop door(s) open to allow some airflow. Many owners use thick litter all year round as it's a great method to keep coops warm during the winter months. Sand, on the other hand, tends to be a better option in warm temperatures. It keeps considerably cooler, has been demonstrated to minimize the risk of disease, promotes healthy feet, and becomes a dust bath instantaneously! Employ "construction sand," also referred to as "river sand" or "washed sand." Play sand is extremely fine and may irritate respiratory passages.Use sand in your chicken coop to offer your hens a day at the beach!Now that you've done these things to keep your hens cool, you can rest and enjoy the warm season with your flock!

Upkeep of a Chicken Coop

A Comprehensive Guide

Imagine this: you wake up to the sound of birds chirping sweetly outside your window, and you find a flourishing chicken coop full of satisfied chickens as soon as you enter your backyard.It makes me pleased to watch them pecking at the ground and enjoying the weather. To ensure that your feathered buddies thrive in a safe and warm location, nevertheless, sustaining a chicken coop calls for labor and knowledge. We'll cover everything, from cleaning measures to predator-proofing strategies. So continue reading if you're willing to construct your treasured flock the mansion of their dreams!

Everyday To-Do List

Daily coop upkeep is vital to the health of your chickens. You may keep the pleasure of your feathered pals by sticking to these uncomplicated measures.

Take Out Uneaten Scraps Every Day

Eliminate any uneaten food, especially vegetable scraps, from the chicken coop as one of the first items on your monthly maintenance checklist. Pests and bacterial growth may be prevented by throwing away leftover leftovers.

Everyday Schedule

Frequently replenish food and water containers.

Like any other pet, chickens have basic needs: they must constantly have access to food and water.To keep them fed and hydrated, it's vital to regularly refresh their food and water containers. Verify that there are no contaminants or debris in the water.

Keep their Bedding Clean

Keeping the bedding in a chicken coop clean is crucial to keeping it healthy. Straw and wood shavings are examples of unclean bedding that should be periodically removed as they may contain germs and generate terrible odors.

To offer your hens a neat and pleasant location to live, replace it with fresh bedding.

Gather Eggs out of Nest Boxes

It's vital to harvest eggs every day if your birds are producing eggs. If eggs are left in the nesting boxes for a lengthy time, they may break or the hens may become more brooding.

Timely egg collecting preserves their integrity and limits the chance of disturbances among the chickens.

Look for Indications of Disease or Injury

Take some time each day to examine your flock for any indicators of sickness or injuries as part of your usual chicken care activities. Keep a watch out for strange behavior, such as changes in the appearance of feathers, drowsiness, or lack of food.It's vital to take prompt action if you observe any troubling indicators to stop the infection from spreading and safeguard your flock.You may offer a clean and healthy environment for your hens by adopting these daily standards for poultry care into your routine.

156

Continual Maintenance

The health of your flock relies on you keeping your chicken coop clean. Maintaining cleanliness in your chicken coop should be your goal, regardless of the cleaning strategy you settle on. Keeping your chickens' environment clean helps reduce the collection of dust, filth, and pathogens that may cause health concerns. Frequent cleaning decreases the chance of infection and sickness by ensuring that any waste or rubbish is immediately cleared. It also assists in the control of unwanted scents that may build up in an unclean coop.

Employ Natural Cleaning Products to Ensure Hygiene

A monthly duty connected to caring for your hens is giving your coop a good cleaning with natural cleansers like vinegar. Because of its antibacterial characteristics, vinegar helps kill germs and bacteria without injuring your chickens.In a spray bottle, add equal parts vinegar and warm water to use as a disinfectant. To get rid of scents and remove germs, spray this solution on walls, perches, and nesting boxes. Do not forget to let the area thoroughly dry before letting the chickens back in.

Dish soap may be used to clean, scour, and wash waterers as required.

Spot-Cleaning versus Deep Littering

Spot-cleaning

Spot cleaning helps you to frequently get rid of any noticeable garbage or unclean bedding. You take out all the bedding once a week and replace it with new pine shavings or straws.

Because it eliminates odors, this is the most widely utilized strategy among backyard chicken owners. It comes highly recommended for chicken coops in Somerzby.

Deep Litter Approach

Using this process, the coop's floor is coated with a thick layer of bedding material (such as wood shavings or straw). Conditions resemble compost as a consequence of the bedding getting polluted over time with droppings and other waste materials.Turning the bedding over frequently aids decomposition.In hot months, this method may be fairly harsh. It also demands taller coop boundaries to prevent compost from overflowing outside the coop. This strategy is not indicated for Somerzby chicken coops.Your hens will live in a healthier environment if you incorporate frequent cleaning in your usual care for the chicken coop. It also makes things easy for you because the mess gets harder to clean up the longer you leave it!

What people needs to know about chicken coops

How Frequently Should My Backyard Chicken Coop Be Cleaned?

It is advisable to undertake spot cleanings every day and to fully clean out all bed sheets at least once every seven days. But, more frequent cleaning might be necessary if you discover any indicators of an unpleasant stink or an excessive collection of rubbish.

What Kind of Food Should I Feed My Chickens To Get Them Healthy?

A well-rounded diet for hens has to contain commercial poultry feed, as well as fresh food, fruits, and grit for simpler digestion. For detailed dietary guidance, talk with a veterinarian.

How Can I Keep Predators Out of My Chickens?

Make sure your coop has a robust fence and wire mesh covering the ground to keep predators out and keep your hens secure. As a deterrent, use motion-activated lights and make sure the coop is closed tight at night.

It's time to put your newfound awareness of the significance of periodic maintenance, cleaning, and thoughtful coop design into practice.Knowing that your coop directly influences the health of your feathered buddies, take joy in keeping it clean and secure. Recall that a pleased hen translates into more scrumptious, fresh eggs for you!

Chapter 6

For your poultry litter, there are many different uses

The poultry industry is one of the industries that is increasing at the fastest rate in the globe. In addition to supplying us with a substantial amount of chicken meat, the company also produces a great deal of waste products.Feathers, spilled feed, sawdust and wood shavings, feces, and feathers. These are the components that makeup poultry litter at the the Farm, where turkeys are kept. This combination contains several nutrients that have extra applications. Some of the factors that influence the quality of chicken litter include the food that the bird consumes, as well as other aspects like the kind of bedding that is used, any supplements that the bird may be given, and how the litter is stored.

Why is it important to reuse the litter from your poultry?

There are several advantages to reusing your chicken litter in a variety of different ways. Reusing your garbage may aid in minimizing the amount of junk that needs to be disposed of. This helps you to save money and time when utilizing your disposal.Poultry waste may be repurposed to boost biosecurity, minimize odor concerns, and remove the need to store vast volumes of waste on-site until they are disposed of. This will aid in arranging things better and cleaning up the home. Both large-scale chicken producers' and small-scale poultry houses' litter may be employed for litter materials.

What additional applications are there for poultry litter?

1. Animal Meals

Chicken litter is a rich source of nourishment for fish and cattle. Cattle benefit from and already absorb many of the nutrients contained in poultry litter. You may substitute poultry litter for cow feed or mix the two to make it more substantial. This can result in cost savings on bulk feed. Poultry litter needs to be thoroughly prepared before it can be given to animals. Feathers and plastic items need to be cleaned out first. Poultry litter may also carry diseases like salmonella. To safeguard the animals that consume it,

microorganisms like these must be effectively eradicated by processing before being utilized as animal feed.

2. Apply as a Soil Conditioner

The optimal fertilizer is made up of poultry litter. It is packed with nutrients that are necessary for plant growth. The three basic constituents of poultry litter are **phosphorus, nitrogen, and potassium.**The feces present in the litter supply the majority of the nutrients in poultry litter. Studies have revealed that soil treated with chicken dung promotes a larger range of crop development. You may cut the expenditure of transportation for the collection of your poultry waste by utilizing your litter as fertilizer.

3. Boost the Quality of the Soil

The quality of soil may be severely degraded and catastrophic harm can ensue from the continual use of chemical fertilizers. A recent study has revealed the positive impacts of chicken dung on the soil's water-holding capacity and structure. The quality and productivity of the crops you grow on your farm will both increase with the application of chicken litter to your soil.

4. Source of Fuel

And finally, chicken poo is an excellent fuel source. It is feasible to burn low-moisture chicken litter to create energy. You may recover heat, hot water, and power for your poultry farm by utilizing one of our W2E (Waste To Energy) systems with your spent chicken litter. One example would be burning the excrement from your chickens to give light to your chicken houses.

Energy waste (W2E): what is it?

Burning rubbish to create energy that may be recovered as heat or electricity is known as a **waste-to-energy** plant.Energy waste may be a particularly valuable approach for farms in emerging and remote places, delivering long-term benefits. Because less rubbish is placed in landfills,the waste-to-energy process helps the environment as well. It is also a technique for optimizing the usage and recycling of waste material. Additionally, it has been stated that W2E may minimize the consequences of climate change. This is because rubbish that is burnt creates fewer dangerous fumes than it would if it were put in a landfill.The most efficient approach to dispose of chicken waste is by **incineration**. High degrees of disease and infection avoidance, total process control, and the ability to recover energy used for power and heat are the consequences.

Poultry Waste Management & Laws

When it comes to treating poultry feces, there are rules and procedures in place. The majority of the limitations are pretty similar, however, they may vary depending on the country you are in.You shouldn't store any poultry waste on your premises for longer than is essential. If a bird is dead, you have to remove it away from the other birds straight away. Keeping dead animals on site raises the potential of infection and sickness as well as biosecurity.

Your litter must be thoroughly handled before removing any toxic or hazardous bacteria if a dead bird is spotted on the litter you wish to use as fertilizer. This may contaminate crops grown on contaminated soil and constitute a health risk to humans and animals who may ingest the products if it is used as fertilizer without being treated.Farmers must make sure that putting chicken litter on their land doesn't cause any stink before it can be recognized as fertilizer. This is especially significant in populous locations.

How to Improve the Eco-Friendliness of Your Chicken Coop

The hobby of rearing chickens is gaining in popularity. Concrete is a common material to utilize for creating chicken coops because of its longevity and toughness. Concrete is hardly the greenest building material, however. Use Recycled Materials When Building a Concrete Chicken Coop for an Eco-Friendly Home: Here are some techniques to decrease the environmental influence of your concrete chicken coop:

Here are some ideas to decrease your concrete chicken coop's harmful environmental effects:

Utilize Recycled Resources

Take into mind utilizing recycled resources when creating a concrete coop.

Concrete aggregate that has been recycled: Utilizing crushed concrete from abandoned structures minimizes the requirement for new concrete. You may use this recycled concrete aggregate for the walls, floors, and foundation.

Reclaimed wood: Reclaimed wood may be used to frame and create fences, nest boxes, and roosts. For salvaged wood, browse via the ads in your region or contact demolition businesses.

Repurposed windows and doors: Install windows and doors that have been saved to bring natural light and air inside the coop. Verify that they have two panes for insulation.

Scrap metal roofing: To keep the coop cool, a robust, reflective metal roof made of scrap material is erected. For protection against corrosion, use galvanized metal.

Selecting recycled building materials minimizes resource consumption, waste, and carbon emissions from the manufacture of new materials.

Put in Insulation

A concrete coop may reduce heat loss in the winter and overheating in the summer by being insulated:Rigid foam insulation boards attached with concrete anchors are put into concrete block walls.For walls built of poured concrete, cover the interior surface of the stiff insulation with plywood.Take precautions to reduce heat leaking through the floor by putting a layer of stiff insulation beneath the slab foundation.To avoid radiant heat, put a radiant barrier on the bottom of the roof.To insulate the ceiling, use blown-in cellulose or fiberglass batts. A ventilation gap should be left along the rooftop.

With the correct insulation, chickens may live comfortably all year long and consume less energy.

Include Solar Energy

One clean, sustainable technique of powering the chicken coop's components is utilizing solar power:Put in place solar panels with a southerly orientation on the roof. Electricity will be generated by this to drive fans, lighting, automatic doors, and other accessories.To provide hot water for cleaning the coop, consider building a solar water heating system with evacuated tube collectors positioned on the roof.For interior light, use a solitary LED bulb that works on solar power. They don't require wires to create brightness.Adding a solar attic fan will boost cooling and air circulation. There are models of thermostats that can be modified.The coop's energy cost and environmental effects are minimized by employing solar power. Benefit from the sun's exposure on the roof.

Gather Rainfall

Large roof areas on chicken coops are good for collecting rainwater: Connect one or more rain barrels to the rain gutters. Gardens or landscapes may be utilized for barrel overflow. Utilize a cistern to retain hundreds of gallons of roof-based rainwater for greater harvests.Gathered rainwater may be utilized for garden irrigation, coop cleaning, and chicken drinking. To eliminate contaminants before rainwater reaches the barrels, use a first flush diverter.Ensure that collecting containers are covered to limit the growth of algae and the breeding of mosquitoes.By collecting rainwater, fewer well water, and municipal water sources are consumed. The fresher taste of chlorinated tap water will appeal to chickens.

Manure Compost and Bedding

Composting is best accomplished with chicken excrement and coop bedding items like straw:Regularly collect filthy bedding and manure into an open pile or compost container. For successful breakdown, stack materials that are brown and green.Through the use of worms, vermicomposting creates nutrient-rich worm castings for plants fast from manure and bedding.Nitrogen,phosphorus,and other critical micronutrients found in compost and worm castings serve to enrich any soil.Compacted compost increases plant growth and soil health. It may be used for potting mixes or worked into garden beds.Manure composting lowers waste and gets rid of bugs and aromas from raw manure piles.Transform your vegetable and flower gardens by utilizing chicken poo to create black gold!

Dim the Building

There are several benefits to encircling the chicken coop with shade trees:Planting deciduous trees on the west and south sides will shelter the coop from the summer heat while permitting winter sunshine.A living roof that is covered with hardy succulents like sedum may increase the coop's insulation and minimize runoff.In the summer, the trellis-growing vines produce lovely shade, while in the winter, they provide insulation. Encircling the perimeter with trees and vines may help keep away the prevailing winds while allowing for ventilation.So that the chickens may enjoy the shade and any falling fruits in their run, grow berry bushes and fruit trees.In addition to providing food and security, shading the coop and its environs will aid in maintaining temperature.

Make Use of Natural Pest Control

Instead of utilizing harsh chemicals to handle pests, there are various natural options available:Aphids, mites, and flies may all be handled by introducing predator insects such as ladybugs or praying mantises.Bed bugs, flies, and beetles lose moisture as a consequence of the waxy coating being worn down by diatomaceous earth powder. Apply gently to the nests and coop floor. After cleaning, reapply.Using kaolin clay, around the coop's foundation. Rodents and snakes that burrow are deterred by the sharp particles. Food-grade DE should be added to dust baths. Using their feathers, hens will distribute them to fight against mites and lice.To keep mosquitoes and crawling insects under control, utilize herbal insect sprays that contain ingredients like garlic, cinnamon, and peppermint oil. Pay attention to hiding areas and access points.By employing natural pest control approaches, groundwater and soil contamination as well as chemical exposure for hens are minimized.

Give robust materials a priority.

Using components that are long-lasting and durable is the ideal technique to lessen waste:Given its extended lifespan, concrete is among the most ecological and durable materials. Cob, stone, and brick houses are additional choices.For high-traffic areas such as exterior pathways, perches, nests, and flooring, utilize cedar or composite decking. Steer wary of pressure-treated wood.Flooring and fence composed of galvanized hardware cloth with 1/2" holes prevent degradation and warping. Another substance that resists corrosion is stainless steel.When properly maintained for, stainless steel waterers and feeders will live for many years. Steer clear of plastic models.

Choosing materials that are long-lasting and durable may initially cost more, but in the long term, it will save money and resources.

Increase Airflow

Enough ventilation is vital for a flock's health:To maximize airflow, orient the coop so that the longer side faces the direction of the prevailing winds. Verify that the apertures line up.Install cupolas, gable vents, and perforations over the roof to enable hot air to escape.Install moveable air vents high up on walls to get rid of moisture and ammonia collection. Summertime expands open vents.To actively remove air, place exhaust fans on the ceilings or ends of gable walls. When it's warmer than 85°F, switch on the fans.Aviary wire covering windows enables ventilation yet keeps predators away. On cool summer nights, open the windows.In the design of the coop, prevent dead air spaces. Airflow is impeded by totally ceiling-high walls and lofts.By minimizing odor, humidity, and overheating, these ventilation strategies offer a healthy atmosphere where hens may grow.

Look into Eco-Friendly Choices

See your local library for books on Creating a Hummingbird Garden and The Eco-Friendly Garden, which will educate you on how to employ native plants for sustainable landscaping around the chicken run and coop. This decreases care while supplying habitat, food supplies, and shade.For examples of ecologically friendly coops with features like living roofs, rainwater collection systems, and solar panels, check out YouTube channels like Tilly's Nest and The Benevolent Chicken. Make notes on prospective designs and material options.For information on how to build up a closed-loop system on your homestead that centers on poultry, check out sites like The Prairie Homestead. Utilize waste and bedding compost to increase gardens, which supply organic fertilizer.Enroll in an online course on current homesteading practices from websites like Udemy. A more sustainable chicken-keeping operation may benefit from the gardening, beekeeping, composting, food preservation, and animal husbandry taught in the courses.Become a member of Facebook groups or local community organizations to meet people who grow backyard chickens and urban farms. The majority of communities have a thriving association that meets often to discuss ideas and general experience on the best methods for rearing chickens.

A robust concrete chicken coop may be designed, constructed, and maintained much more readily with the assistance of these excellent online resources.

WAYS TO IMPROVE YOUR POULTRY FARMING EARNINGS

You will undoubtedly need to think about ways to enhance profitability in your firm. Making a profit is not simply something that novices do. Even seasoned chicken farmers are seeking inventive methods to enhance their production and income from the poultry sector. Who doesn't want to enhance their income, after all?HAHA!These recommendations for Increasing Your Earnings in Poultry Farming will be beneficial given the present high cost of inputs for poultry farms. You'll discover valuable tips in this section to aid you in optimizing your poultry farm's operations and raising earnings.

Advice on How to Boost Your Income in a Poultry Farming Business

You may be able to accomplish so without raising the cost of your animals. Enhancing your present procedures could offer you with a competitive advantage over other chicken growers. Allow me to explain it to you. The process of producing chickens involves numerous expenses. If you can effectively optimize the cost of producing your chickens, you will continue to earn money even while others suffer. What then can you do to enhance the earnings from your chicken business?

Start by picking the greatest breeds of chickens.

Getting this one properly is vital if you want to optimize revenues from your poultry company. This is true because, as a chicken farmer, everything you do is centered on this one option. You will most likely lose a lot of money in the future if you start with a substandard breed of chicken.Having the broadest choice of chicken breeds is the first step in sparing you from this issue.Now, the difficulty is how to pick the finest breeds of chickens for your poultry company. Here are some steps you may perform to aid you in deciding.

Establish the purpose of your chicken company.

This is crucial as it will help you pick the right breed for farming. First, you have to select if you want to participate in broiler farming, or the production of meat. You may now pick the ideal breed of broiler chicken for you in this scenario.Pick a broiler chicken that will mature to table-size weight in roughly 8 weeks. On the other hand, you have to pick from a list of layer chickens if you intend to build a farm or produce eggs.Select only very productive layer breeds that are resistant to illnesses.You should contact a veterinarian or ask other farmers on online forums if you are uncertain about what to do.

Make your poultry feed better.

Another crucial feature that could help you enhance the revenue from your chicken farm is feed optimization. Feed takes up a major percentage of the cost of raising hens, hence feed optimization is crucial. The birds also require water and food, which they absorb into their bodies to make eggs and meat. To optimize your poultry profit, you need to cut down on the expense of feeding the birds.Adding supplementary feeds to their diet, such as fodder and black army fly (BSF), is one effective approach to enhance their nutrition.For additional sustenance for hens, you may cultivate black army flies or plant fodder. To optimize income, you may also invest your money in a regular feed formula for your hens,to employ organic treatment as opposed to synthetic.The price of medicine is another component that greatly adds to the cost of raising hens, just behind the cost of food.For their birds, many farmers spend a lot of money on synthetic treatments. You would have spent a lot of money and generated very little profit after the manufacturing phase.Using organic medicine is a proven technique to stay clear of this and boost your income. Herbs and spices used in organic chicken treatment serve to improve the immune system and prevent poultry disease.By choosing organic alternatives, you may now remove all of those fake treatments and enhance your earnings. You may boost your profits from poultry farming and lessen chicken mortality with the aid of these organic remedies.

Build robust buildings.

A large percentage of the expenditures connected with launching a viable poultry farming enterprise go into constructing a **chicken coop.** You must consequently take great care to guarantee that this is completed once and for all. You simply need to do it well. When constructing your chicken coop, always choose the best materials, regardless of whether you decide to utilize the battery cage technique or the deep litter approach. A robust structure will increase your chicken revenue and save you money on maintenance.

In summary

You must put the four above tips into practice if you're a poultry farmer aiming to maximize revenues from your birds. They will aid you in minimizing needless expenditure and put you up for better financial success in your poultry business. Putting these methods into practice will, in my view, drastically change your game. Please help them boost their revenue by passing this article to everyone you know.

Chapter 7

Beginner-Friendly Online Forums and Resources

Making the most of the quantity of knowledge available in online forums and other resources is vital for newbies searching for guidance, advise, and a sense of belonging in the ever-changing world of hen care. The internet offers a broad variety of platforms on which beginner chicken keepers may engage with more seasoned enthusiasts, share ideas, and have access to a plethora of beneficial material.Participating in online chicken-keeping forums allows newbies a virtual location to ask questions, tell tales, and gain help from a community of individuals who share their interests. There is a sense of camaraderie among chicken owners owing to forums maintained by agricultural extension offices, sites like **BackYardChickens**, and The **Poultry Site.** These discussion forums are highly useful for addressing difficulties, discussing best practices, and interacting with other poultry enthusiasts.

Social media platforms are also significant for the growth of online communities dedicated to hen keeping.

Facebook groups, Reddit groups, Instagram groups, and other social media platforms serve a broad range of hen keepers, from suburban backyard hobbyists to rural homesteaders. These sections allow users to connect in real time, sharing photographs, asking for quick advice, and taking part in conversations about various parts of keeping chickens.For beginners, vast information hubs are supplied by specialized poultry websites and internet resources picked by reliable organizations and professionals. There are a ton of articles, recommendations, and research-based resources accessible on websites like the Livestock Conservancy, the American Poultry Association, and the University of Florida's IFAS Extension. These books equip newcomers with a firm foundation of knowledge by covering everything from breed selection and coop construction to health management and sustainable practices.Beginners' learning is further boosted via poultry-keeping-specific podcasts and webinars. Expert comments from seasoned specialists and seasoned chicken keepers may be obtained on audio platforms such as The Chicken Thistle Farm CoopCast and webinars provided by agricultural institutions and poultry clubs. For those who are just beginning, these formats give a simple method to learn while doing other things or giving your hens some hands-on care.For beginners who seek a more comprehensive and systematic approach, online courses particularly created for rearing chickens offer an organized learning environment. Online learning settings such as Coursera, Udemy, and local agricultural extension offices give courses on a range of disciplines, from biosecurity and breeding to basic chicken care. The objective of these courses is to give beginners important knowledge and skills to enhance their experience with rearing hens.Online courses, podcasts, social media groups, forums, and specialist websites all add to a rich ecosystem of resources for beginners entering the realm of hen rearing. For persons just beginning in chicken husbandry, the ease of access to information and the possibility to connect with a broad spectrum of enthusiasts online considerably minimize the learning curve. Accepting these internet resources puts beginners in a better position to make informed decisions, handle difficulties, and have a happy and long-lasting time in the henhouse.

Taking Part in Events and Poultry Shows

Poultry enthusiasts regard visiting poultry exhibits and events as a rare opportunity to display the fruits of their hard work, exchange experiences, and become part of a dynamic community. From regional fairs to specialty displays, these events offer a place for chicken keepers to celebrate their passion and give back to the greater poultry community.

A significant benefit of joining a poultry show is having the ability to present your flock and gain valuable critique from qualified judges. These evaluations not only affirm the effort put forth in raising wholesome, well-bred chickens, but they also give important information for continued improvement. Judges examine numerous variables, such as overall health, plumage quality, and compliance with breed standards, which inspire contestants to strive for excellence.One evident advantage of participation in poultry shows and events is networking. These get-togethers attract a broad range of poultry enthusiasts, from expert breeders to newbies with lots of questions. In these sorts of environments, individuals exchange knowledge by discussing experiences, recommendations, and best practices. These events typically result in talks that go beyond the exhibition floor and develop lasting ties that serve as the cornerstone of a helpful poultry community.In addition to offering a forum for exhibitors to present their birds, poultry shows also give viewers with a chance to learn more about different breeds and husbandry practices. Exhibitions frequently feature presentations on poultry health, information specific to a particular breed, and expert demonstrations. This thorough training program delivers a profusion of knowledge to newbies to enhance their experience with hen rearing.Poultry exhibits are not just for seasoned breeders to attend; beginners may learn a great deal from them as well. Several contests divide participants according to ability levels so that novices may participate with their peers. With this all-inclusive approach, newbies are encouraged to take center stage, obtain constructive feedback for their work, and progressively improve their poultry-keeping methods.Poultry exhibits generally feature entertainment and activities for both families and enthusiasts, in addition to the competitive components. Attendees are pulled in by the interactive demonstrations, poultry-themed items, and informative displays. These components contribute to the overall attraction of poultry shows, making them popular events for spectators as well as contestants.Apart from local contests, regional and national poultry events give a larger platform for individuals desiring to raise their level of engagement. Events like the annual National Meet of the American Poultry Association bring together poultry connoisseurs from all around the country and generate a sense of solidarity among like-minded people. Attending such events provides the opportunity to witness a broad range of breeds, pick the brains of famous breeders, and become completely immersed in the larger world of poultry culture.

To sum up, taking part in poultry competitions and exhibits is a complicated experience that transcends beyond basic competitiveness. It offers a place for community growth, learning, networking, and validation. Taking part in these activities, whether you are an experienced breeder or a novice enthusiast, makes the poultry-keeping community more active and energetic by providing a common space where the love of hens is cultivated and celebrated.

In summary

After "Novice's Handbook to Chicken Keeping," it's evident that maintaining satisfied and healthy backyard chickens is a science as well as an art. This booklet offers beginners a way to comfortably navigate the complexity of poultry care with a plethora of practical ideas and tried-and-true strategies.Every chapter enhances the entire development of a beginner chicken keeper, from selecting the greatest breeds of hens and constructing the right coop to grasping local rules and creating ties with poultry doctors. A compassionate approach mixed with a focus on practicality underlines how crucial it is to offer a quiet atmosphere for both caregivers and their feathered companions.This manual instills a sense of responsibility and stewardship in newcomers while providing them with the required knowledge for efficient chicken care. It achieves this by investigating the nuances of chicken behavior, health management, and sustainable ways. It's about building a symbiotic relationship with these lovely creatures, enhancing the lives of both the keeper and the kept, rather than merely breeding hens.Novices learn the capacity to solve issues, make smart decisions, and actively interact in the greater poultry community as they work their way through the chapters. The handbook offers more than simply a guide for anybody beginning the pleasant journey of rearing hens. With the aid of this resource, novices may create a rewarding, long-lasting, and compassionate friendship with their backyard flock.The *"Novice's Handbook to Chicken Keeping"* is more than simply a practical handbook; at its heart, it's a philosophy of care and concern for these feathered creatures. This handbook intends to serve as a guide for novices, giving a clear route to a satisfying and instructive experience in the world of backyard chicken keeping by combining valuable information with tried-and-true strategies. I hope that the information presented in these pages will help to develop a thriving community of responsible chicken owners who are devoted to the health and happiness of their feathered friends.